Contemporary Discourse in the Field of BIOLOGY™

The Applications and Limitations of Taxonomy (in Classification of Organisms)

An Anthology of Current Thought

Edited by Jeri Freedman

The Rosen Publishing Group, Inc., New York

Published in 2006 by The Rosen Publishing Group, Inc.
29 East 21st Street, New York, NY 10010

First Edition

Library of Congress Cataloging-in-Publication Data

The applications and limitations of taxonomy (in classification of organisms) : an anthology of current thought / edited by Jeri Freedman.
p. cm. — (Contemporary discourse in the field of biology)
Includes index.
ISBN 1-4042-0400-8 (library binding)
1. Biology—Classification.
I. Freedman, Jeri. II. Series.
QH83.A785 2005
570'.1'2—dc22

2004021800

Manufactured in the United States of America

On the cover: Bottom right: Historical artwork showing the taxonomic classification of animal species according to Baron Georges Cuvier. Top: Digital cell. Far left: Digital cell. Bottom left: Austrian monk and botanist Gregor Johann Mendel (1822–1884).

CONTENTS

Introduction

Taxonomy, or systematics as the field is sometimes referred to, is the classifying of organisms in a hierarchy that reflects their relationships. Classifying organisms not only sheds light on the relationships among plants and animals, but it also facilitates our understanding of their origins.

Attempts to classify living things date back to the *History of Animals* by Aristotle (384–322 BC) and *De Plantis* by Theophrastus (c. 381–276 BC). Such works provided the basis for scholars' knowledge of plants and animals throughout the Middle Ages. However, these works only dealt with a small number of existing organisms.

Our modern classification system dates back to Carolus Linnaeus (1707–1778), an eighteenth-century Swedish naturalist. Linnaeus developed the first systematic classification system for plants. His system categorized them as unique species that were then grouped into progressively larger and related groups called genera, orders, classes, and kingdoms. In 1735, Linnaeus published the first edition of his classification

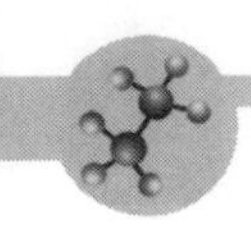

system in a volume called *Systema Naturae* (The System of Nature). Linnaeus's system of classification—which is still in use today (for animals as well as plants)—focuses on classifying organisms according to their physical traits, such as how many petals a flower has.

However, just where organisms fit into a hierarchy is not always obvious. Whales and fish both have fins and swim, but they are not members of the same species—nor are they even closely related. Furthermore, not all biological and genetic characteristics that organisms have in common are easily identifiable. Many require advanced technology to uncover them. The picture becomes even more complicated when one attempts to track the evolutionary relationships among organisms. As plants and animals evolved, their physical characteristics changed. This means that evolutionary relationships—such as that between dinosaurs and birds—may not be apparent from simply examining the physical attributes of organisms. Ultimately, however, more precise methods of classifying organisms were needed. And, as in many areas of modern science, technology has provided new tools.

Taxonomy is experiencing a technology boom that is radically altering the landscape of classification. In the near future, it is likely that species will no longer be grouped into families merely on the basis of their physical characteristics. DNA analysis and other techniques of molecular biology are allowing scientists to identify relationships that are not apparent from merely examining an

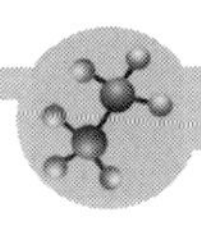

organism's physical traits. Advances in three-dimensional, or 3-D, imaging technology have allowed a comparison of the structure of fossils and bones at a level that was previously impossible to ascertain. In addition, great advances in computer hardware and software are opening up the possibility of creating a universal tree of life in which all the species on Earth are placed in proper relationship to each other on one comprehensive family tree.

However—and as often is the case—along with new techniques has come controversy. Just what are the best and most accurate technologies for identifying genetic and molecular characteristics of animals? What hierarchical method will most clearly show the relationships among plants and animals? In the past, organisms were identified as bearing a resemblance to a representative type of specimen and arranged in a hierarchy of similar organisms based on the possession of common characteristics. One new naming system called PhyloCode would be based more explicitly on evolutionary relationships. Instead of being grouped into ranks, such as genus, family, and order, organisms would be assembled into "clades," which are defined as any set of organisms with a common ancestor. Another approach relies on finding species-specific gene sequences that would act as unique identifiers of each species—much the same as, for example, what bar codes do for products in the supermarket.

Another question is, exactly where do certain organisms fit on the evolutionary tree of life? Not only is it difficult to establish the relationships between

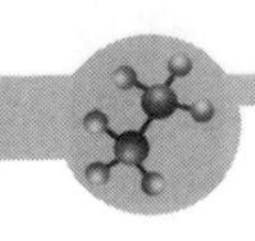

disparate plants and animals, but the exact sequence of human evolution is still unclear. Who exactly were the ancestors of modern human beings and what are the biological and evolutionary relationships of these various prehistoric people?

The articles in this book have been selected from a range of scientific journals. They were chosen with the aim of providing a comprehensive overview (including cutting-edge discourse) of the most significant questions in the field of taxonomy, or the classification of organisms. These questions range from such broad concepts as what is the best technique to use to classify organisms, to specific species-related queries such as whether dinosaurs were warm- or cold-blooded, and just what does the evolutionary tree of modern human beings look like?

This book is divided into six chapters. The first chapter, "What Is Taxonomy?" explores basic questions that are key to the field today, such as "what is the best way to classify organisms?" Currently, many competing approaches are being proposed—each with its advantages and disadvantages. The second chapter, "The Problems of Classifying Plants and Animals," explores some of the confusion and difficulties scientists face in trying to classify specific organisms and the new approaches that are currently being explored. This includes the ranking of organisms on an evolutionary hierarchy and the use of cladistics to classify organisms using a variety of characteristics beyond mere observation of similar physical characteristics. The third chapter,

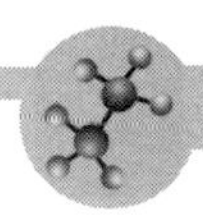

"DNA Analysis and Taxonomy," explores the use of recent DNA sequencing technology to uniquely identify plants and animals, and looks at how DNA analysis is revealing the surprising relationships between apparently disparate organisms. Chapter 4, "Discovering the Truth About Dinosaurs," discusses recent information that challenges what we thought we knew about dinosaurs. This includes a discussion of new information and scientific techniques that are raising questions about how to classify dinosaurs. In addition, recent discoveries are causing a reevaluation of the place of dinosaurs in the evolution of other species such as birds. The fifth chapter, "Who Were Our Ancestors?" explores the problems encountered in trying to establish the evolutionary tree that starts with prehistoric proto-humans and leads to modern human beings. And the final chapter, "Can We Create a Universal Tree of Life?" examines this question by reviewing the work under way with two different efforts using different technologies, the difficulties that must be overcome to achieve such an ambitious project, and the controversies around these efforts.

We hope this book will shed some light on why it is important to classify organisms and what we hope to learn by doing so. In addition, it should provide some insight to the major questions faced today in attempting to understand living organisms, and to some of the new techniques that are being used to further our understanding of the natural world. *—JF*

1 What Is Taxonomy?

The first article, "The Name Game," looks at the language used for classification. As with many areas of scientific study, the field of classifying plants and animals, called taxonomy, has its own technical vocabulary. International standards for naming plants and animals have been developed so that scientists have a consistent way of referring to organisms. This universal language makes it easier to understand what is being referred to regardless of who is speaking or writing about an organism or where he or she is located.

Given the vast number of plants and animals on Earth, there would be chaos and confusion when trying to figure out which organism was being referred to if there were no consensus on the terminology being used. The following article provides a useful introduction to the terminology of taxonomy, an explanation of the problems that the process of naming organisms evolved to address, and

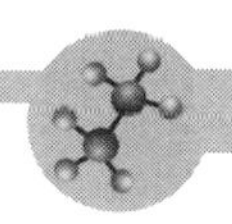

the reasons that certain naming conventions are used. —JF

"The Name Game: A Few Words About Understanding the Language of the Natural World"

by David Petersen
***Backpacker*, April 1993**

A few words about understanding the language of the natural world.

Studying the natural world—even if it's in your own backyard—requires proficiency in a foreign language, the language of wildlife biologists. Like doctors and lawyers, these scientists have their own way of saying things, and although it may sound like mumbo jumbo to most people, the verbiage can actually help you understand the complexities of Mother Nature. For instance, in light conversation one biologist might ask another, "Did you happen to note the circumference of the preorbital glands on that new *Cervus elaphus* mount?" To which his companion could reply, "Not yet, for I have been much too busy measuring the posterior spinal appendages of the *Onychomys leucogaster* specimens, comparing them against *O. torridus*."

Indeed.

You've probably wondered, while browsing through a field guide to plants or animals, why scientists insist on using tongue-twisting foreign words to identify plants and animals. Why can't they just call an elk an elk?

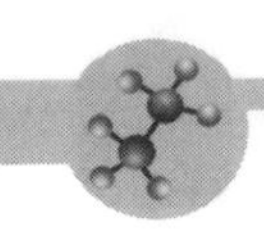

Upon that question hangs a posterior spinal appendage.

The branch of biological science charged with identifying and naming species of living things is called systematics, or more commonly, taxonomy. With known species going extinct faster than new ones are being discovered, it's not such a big job now, but prior to the mid-1700s, taxonomics was a mess. Since there was no universally accepted system of scientific nomenclature in place then, each time someone discovered a new species it was named essentially by caprice, usually in whatever tongue the discoverer happened to speak.

Consequently, two or more scientists working independently often discovered an identical new plant or animal. But with each being ignorant of the other's discovery, or language, they assigned redundant and conflicting names.

With more biologists hitting the fields every year and more new species being discovered and named, and renamed, this identification mayhem just couldn't be allowed to continue. Finally, in 1758, the Swedish botanist Carolus Linnaeus introduced the "binomial nomenclature" system of biological taxonomy. It was quickly accepted as the system of biological taxonomy and remains the standard today.

As its name implies, the binomial nomenclature system employs two words to describe a given life form. These words are generally drawn from Latin, or they may be latinized English words or names from other languages, or, less commonly, they may be from ancient Greek. The first of the two words is italicized

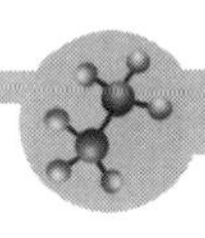

and capitalized and identifies the plant or animal's genus, or sub-family (*Homo*, for example, which means "man"). The second word, also written in italics but not capitalized, is the species identifier, or "specific epithet" (e.g. *sapiens*, which means "wise").

Beyond that, it gets more complicated. Frequently, a species is further divided into two or more subspecies. White-tailed deer are classified within the genus Odocoileus and the species virginianus. But under that specific epithet fall some thirty subspecies. One of these, by way of example, is the tiny Coues deer of the southern Arizona and New Mexico deserts. To distinguish the Coues from the twenty-nine other subspecies of whitetails, a scientist would say, or write, *O.v. couesi*.

The binominal nomenclature system worked smoothly for the remainder of its inventor's life and then some. But before another century had passed, some 400,000 new species had been discovered and more were turning up every day, harshly testing the Linnaean approach. One of the primary bones of contention was how to determine who had naming rights when a new species was discovered virtually simultaneously by multiple researchers. Linnaeus had outlined how a species was to be named, and in what language(s), but offered no guidance as to who had the right to do that naming.

The problem was ultimately resolved in 1889, when a group of scientists meeting in Holland formed the International Congress of Zoology. Out of that congress was born the International Commission on Zoological Nomenclature (ICZN), whose formidable responsibility was to establish contentment amid contention.

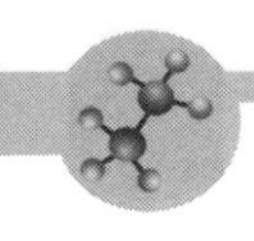

To resolve the naming wars, the ICZN instituted the "law of priority," which states that the first name assigned to a species under tile Linnaean system has priority over later names. Generally, that means the first scientist to discover and name a new species, and publish that discovery and name, has it in the bag.

But you may be wondering why scientists still employ two dead languages to describe life forms rather than, say, Linnaeus' native tongue. Well, simply because Greek and Latin are foreign to everybody. This avoids international jealousies certain to arise if favoritism were shown toward any living language or languages, as well as linguistic confusion if names were assigned in a hodgepodge of tongues. But more important, I suspect, is the fact that scientists, like doctors and lawyers, like to keep the rest of us guessing.

Although Charles Darwin, in his book On the Origin of Species, *provides evidence that species evolve over time, it remains an open question. What causes new species to emerge?*

There are two very noticeable trends in recent theories proposed to account for the differentiation of species. There has been a shift of focus away from theories that emphasize speciation, or the process of biological species formation, as the

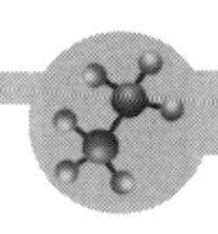

direct result of significant environmental and geographic differences. Instead, there is a new focus on situations where speciation takes place without any dramatic upheaval in a population of similar creatures in the same environment. The second trend is an increasing use of mathematical models to study the phenomena of speciation. Applying these models to the populations of similar animals in like environments has produced some interesting results, as discussed in the following article. —JF

"How the Species Became: Did the Phenomenon Responsible for Sand Dunes and Magnets Also Help Create Everything from Earwigs to Elephants?"

by Ian Stewart
***New Scientist*, October 11, 2003**

One of the ironies of Charles Darwin's *On the Origin of Species* is that while it provides ample evidence that new species evolve from existing ones, it doesn't tell us much about how it happens. It is easy to see that natural selection can cause a species to change as time passes, but it is much less clear why a single species should split into two distinct branches of the evolutionary tree. If some external change makes certain members of a population more able to survive than others, then surely that change will make the whole species evolve in that direction. How could two separate species emerge from one?

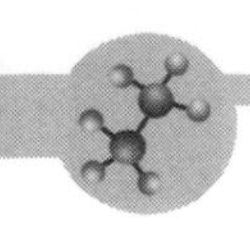

Speciation is a complex business, taking place over vastly different scales of size and time. There is no reason to suppose that it is governed by just one force—after all, we know that genetic mutations and sexual recombination of existing genes vie with environmental influences, depletion of resources, parasites, migration and disease. But although many theories and ideas have been offered up to account for speciation, it remains one of the big puzzles in biology.

Within the flurry of activity that surrounds this conundrum, there are two very noticeable trends. One is a shift of focus away from theories in which species formation occurs as the direct result of major environmental and geographic differences. The new focus is on situations where speciation takes place without any dramatic changes, in a single interbreeding population of very similar creatures, all in much the same environment. Nearly all publications on speciation in Nature and Science over the past five years or so focus on this undramatic scenario—a complete reversal of what used to be the case.

The other trend is the growing use of mathematical models, a technique more usually employed to explore aspects of physics. These models are being used by myself and others to describe the natural dynamics of speciation. And when applied to the "undramatic" instances of speciation, they have produced some very interesting results.

The maths indicates that far from being a surprising phenomenon, it would be very odd if speciation didn't occur. It appears to be a result of exactly the same

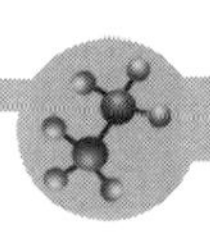

process that filled the universe with matter, creating subatomic particles, planets, sand dunes and—ultimately—humans. Strange as it may seem, neutrons, narwhals, electrons and elephants in some way seem to owe their diverse characteristics to a principle that dictates much of what happens in the physical world.

That principle is known to physicists and mathematicians as "symmetry-breaking." An example is the formation of sand dunes. Reducing it to its mathematical ideal, a uniform wind blowing across a uniform desert will produce parallel ridges of sand. The featureless desert had all the symmetries of a flat plane: rotate it through any angle and it will look the same. The wind, however, reduces the level of symmetry: the parallel ridges of the dunes introduce a definite direction, or orientation, into the landscape.

Such symmetry-breaking happens naturally all over the place. For example, if you heat a flat dish of fluid uniformly from below, at a certain critical temperature the uniformity is broken by the onset of a complex pattern of convection. They are typically hexagonal, with a few pentagons thrown in, and all much the same size. As with the formation of sand dunes, the symmetry breaks down, in this case reducing to the symmetries of a roughly hexagonal lattice.

Filling the Universe

On a far grander scale, physicists believe a type of symmetry-breaking was responsible for the formation of subatomic particles from the fields that filled the primordial universe. These particles are, of course, the

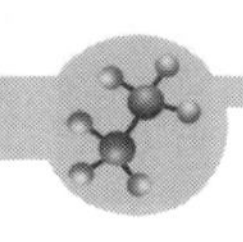

building blocks of matter, so you could argue that symmetry-breaking helped create pretty much everything that exists.

But what does symmetry-breaking have to do with speciation? Although the commonest definition of "same species" in sexual populations has been "able to interbreed," biologists have been seeking an alternative definition for some time, because there are too many cases in which this one just doesn't fit. In a paper published in *BioEssays* this year (vol 25, p 596), Massimo Pigliucci of the University of Tennessee in Knoxville analysed nine well-known definitions of "species," and found serious problems with them all.

So instead of chasing a formal definition of species, biologists are going back to the more intuitive idea that organisms belong to the same species if they are effectively indistinguishable. The degree of similarity can be quantified by listing anatomical or behavioural features and observing how closely they match. And this is where symmetry comes in.

The symmetry of an object or system is simply a transformation that preserves its structure. With speciation, the transformations are not rotations or flips, as with the symmetries of a sphere or a hexagon, but permutations—shufflings of the labels employed in the model to identify the individual organisms.

A group of ten identical objects possesses a symmetry: line them up and turn your back for a moment, and you wouldn't know if any or all of them switched position in the line. But if the line were composed of five objects that had one shape followed by five that had another,

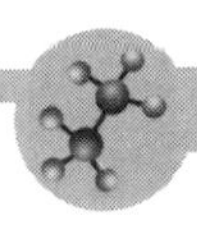

some of the symmetry is broken: swapping numbers 5 and 6 around, for instance, would produce an obvious change.

From this point of view, the definition of a species is simply that it is symmetric, and speciation is then just a form of symmetry-breaking. With this definition in place, mathematicians and physicists can apply their existing theory of symmetry-breaking. This describes how, why and when a given group of symmetries will typically break up into subgroups—species, in this case.

Biologists traditionally recognise two distinct types of speciation. The first is "allopatric" ("different family") speciation, in which some major geographical change splits a population in two. Once separated, the two groups evolve independently, eventually changing so much that they become two distinct species. Even if reunited, they remain distinct species.

The second is "sympatric," or same family speciation, in which new species emerge without separation. You might think that mating between neighbouring animals would encourage "gene flow"—the mixing of "alleles" or gene alternatives that occurs when individuals mate—and would tend to keep the gene pool homogeneous. Classically, this was interpreted as keeping it a single species. But it seems this isn't always the case. Examples include the recent discovery that there are two species of African elephant, and the 13 species of finch on the Galapagos Islands, which helped set Darwin on the road to what he called "the mutability of species."

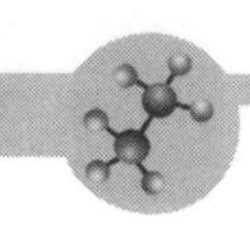

Until around the mid-1990s, allopatric speciation was thought to be by far the most common, but biologists now seem to have begun shifting their view: sympatric speciation, though subtle and counter intuitive, may be the more important mechanism. The homogenising gene flow within a species can be disrupted by many things: geography is just the most obvious. And that's exactly what the mathematical analysis of speciation seems to be suggesting.

The mathematical picture of speciation highlights at least three "universal" phenomena—rules, almost. The first is that when a population first specialties, it usually splits into precisely two distinguishable types. To see three or more new species is a rare, mainly transitory phenomenon. The second is that the split occurs very rapidly in the population—much faster than the usual rate of noticeable changes in characteristics, or phenotype. So, for example, a significant change in beak length might happen within a few generations, rather than by tiny increments over many generations. The third phenomenon is that the two new species will evolve in opposite directions: if one evolves larger beaks, the other will evolve smaller ones.

So what causes that initial split? Our knowledge of symmetry-breaking in physics suggests that a key step is the onset of some kind of instability in the population. An example in physics would be a stick being bent by stronger and stronger forces: something suddenly gives way and the stick snaps in two. Why? Because the two-part state is stable, whereas one over-stressed stick is not. The loss of symmetry is rapid—and irreversible.

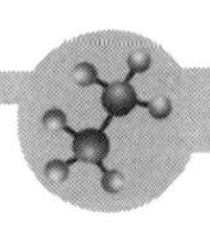

The symmetry-breaking models for speciation do indeed indicate that instability can be a trigger. To be precise about it, a species is called "stable" if small changes in form or behaviour tend to be damped out in subsequent generations. It is unstable if they grow out of control as new generations shuffle the genes of their parents and natural selection discards combinations that don't work so well.

Speciation models show that if you subject a population to subtle, gradual changes in environmental or population pressures, it can suddenly cross a threshold from stable to unstable. At that point, all hell breaks loose. As the environment or population size changes, the single-species state may cease to be stable, so that if by chance a few birds diverge from the average phenotype, the divergence grows instead of damping down. The result is that small, random disturbances can lead to big changes.

Thanks to tiny, random variations that occur naturally—in, say, the beak sizes of a population of finches—anything that changes the characteristics of the food supply even slightly can be an advantage on birds with slightly above or below average beak size. The mathematical analysis shows that once the balance swings in favour of avoiding the middle ground, there is a collective pressure that rapidly drives the birds into two distinct types that don't compete directly for food. Instead, they avoid competition by exploiting distinct niches. This demolishes the argument that the whole population ought to evolve in the same direction, and it opens the door to species divergence in a uniform, interbreeding population.

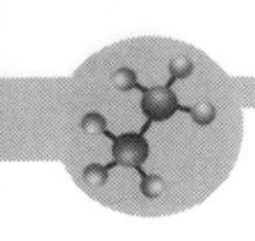

Either of these clumps may split again later, as continuing changes to the environment change the availability of resources. Such a sequence of sympatric splittings is probably how the original single finch species on the Galapagos Islands became 13 (or 14, counting one further species on the Cocos Islands).

The model I have talked about is highly simplified: all creatures in the given species are identical. But researchers are working to remove this crude approximation. One approach is simply to add "random noise" to the equations, so that the phenotype (body-plan and behaviour) passes from generation to generation as per the original rules, but plus or minus small random variations. In this scenario, a population corresponds to a cluster of organisms in phenotypic "space"—an abstract space whose "coordinates" might be beak size, wing span, and so on—rather than a single point. Interestingly, we have found that the clumps split up in much the same way that the original points divided, but the noise makes speciation happen a little more easily.

An even more intriguing approach is to use the original noise-free equations but to modify them so that at each generation, the creatures mate according to a set of randomly chosen pairings. In this scenario, all variation is caused by mating. Again, what we see is clusters, not single points, but these clumps are tighter than in the random noise model, since the homogenising flow pulls the population closer together, just as Harvard zoologist Ernst Mayr has argued it should.

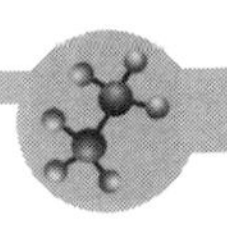

Birds of a Feather

But contrary to what Mayr thought, divergent splits can still occur, and when they do, a tight clump diverges into two much looser ones. After that, even though the creatures can choose mates from the other group, the clusters tighten up and remain separate from each other. This behaviour is not fully understood even mathematically (though curiously, it appears to be related to fractal geometry), but it seems very close to what happens in real populations. All the same, we can see that sympatric speciation is not as surprising as we first thought.

This kind of modelling is still in its infancy. Its main achievement so far is to show that sympatric speciation is entirely reasonable and natural, and to focus attention on the role of instability as a mechanism for speciation. Species become unstable when small but critical changes to their common environment—what food is available, for example—make new gene combinations superior to the existing ones. This alone can cause the phenotypes to rapidly diverge, and contrary to what many textbooks say, it doesn't require mutations within the DNA code. The population just shuffles its existing genes into new arrangements.

Work is already under way to make the models more biologically realistic in order to give a deeper understanding of the nature of these instabilities. My University of Warwick colleague, mathematician Toby Elmhirst, has modelled finch speciation, for instance.

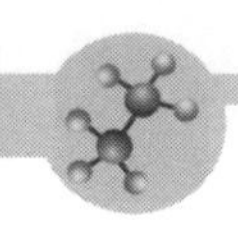

His research follows work done in 1999 by Ulf Dieckmann of the international institute for Applied Systems Analysis in Laxenburg, Austria, and Michael Doebeli of the University of Basel. The approach may tell us about the effect of non-uniform habitat—allopatry as well as sympatry, or subtle mixtures like patchy environments—and it can model sexual and asexual reproduction. The result may yield new insights into how evolution has occurred in the past by placing more emphasis on the links between phenotype and habitat.

Another major objective is to incorporate genetics explicitly. At the moment, its role is implicit: it simply lets the phenotype change. It would be a huge achievement to establish the link between the detailed genetic changes and the phenotypic ones. Such breakthroughs may come from work being carried out by other groups, such as those of Eva Kisdi and Stefan Geritz of the University of Turku in Finland. Their method, known as "adaptive dynamics," follows the same general line of thought as ours, but uses markedly different models, it focuses on the genes in very much the same way we focus on phenotype, and their results give insights into the forces behind allopatric speciation (*Evolution*, vol 53, p 993). The two approaches seem very complementary, and I hope (and expect) that they will join forces as the subject unfolds.

In the meantime, we can at least say that the symmetry breaking approach puts the whole problem in a new light. Species diverge because of an unmanageable

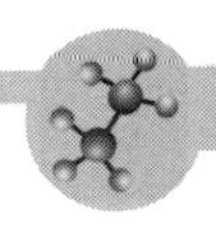

loss of stability. The actual sequence of events—which gene does what, and in what order—determines the precise response to this loss of stability, but it depends on a bewildering variety of accidental factors, such as which birds get the bigger beaks and which get the smaller ones. Broadly speaking, such details are less important than the overall context. They may appear to be the causes of speciation, but actually, they are just the effects of a far-reaching instability. An over-stressed stick must break. An over-stressed group of birds must either speciate or die. Speciation is not surprising—it is simply how the world works.

You might think that the process of classifying organisms is straightforward. Pick out a few common criteria or look for similarities in various plants' or animals' DNA and pop them into neat little boxes. However, the process of categorizing organisms is not that simple. In fact, there is much debate in the scientific community about what criteria should be used to classify plants and animals.

Some taxonomists have been arguing in favor of replacing the traditional system (that was developed by Linnaeus) for describing

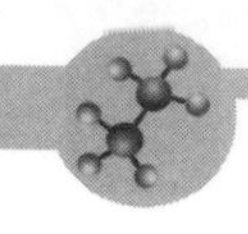

organisms with a Latin genus and species name and organizing them into groups based on how closely they resemble one another. The proponents of one such approach have suggested a new system that they claim is better given our present knowledge of evolution. This new system, called PhyloCode, organizes organisms according to how closely they are related on the evolutionary tree of life.

The following article explains the nature of PhyloCode as well as its advantages and disadvantages as compared to the traditional Linnaean approach to classification that is currently in use. —JF

"Linnaeus's Last Stand? (Introducing PhyloCode)"

by Elizabeth Pennissi
***Science*, March 23, 2001**

A fight has erupted over the best way to name and classify organisms in light of current understanding of evolution and biodiversity

These days the once-serene hallways of the world's natural history museums are anything but tranquil. A small but powerful contingent of systematists is challenging more than 2 centuries of taxonomic tradition by proposing a new system for naming and classifying life, one they say is more in line with the current understanding

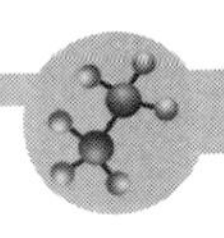

of evolution. Their brash proposal, which will be debated at a symposium in Washington, D.C., on 30 and 31 March, has raised the ire of the more conservative leaders in the field. The resulting controversy over the new naming system, known as "PhyloCode," has pitted colleague against colleague, office mate against office mate. "You've got people willing to throw down their lives on both sides," says Michael Donoghue, a phylogenetic systematist at Yale University in New Haven, Connecticut.

Although few biologists pay rapt attention to systematics, the new proposal, if it prevails, could broadly affect how people think about the biological world. For more than 200 years, a Latin "first" and "last" name—genus and species—has been de rigueur for each organism on Earth. No matter what a person's native tongue or the common name of a species, "Quercus alba" identifies the same exact tree species—white oak—the world over. Yet under PhyloCode, which seeks to reflect phylogenetic relationships, genus names might be lost and species names might be shortened, hyphenated with their former genus designation, or given a numeric designation. The critics are not happy.

The traditional system groups organisms in part according to their resemblance to a representative "type" specimen and places them in a hierarchy of ever more inclusive categories called ranks that have helped people organize and communicate their thinking about flora and fauna. The new naming system would be based more explicitly on evolutionary relationships. Instead of being grouped into ranks, such as genus, family, and order,

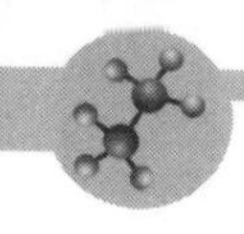

organisms would be assembled into "clades," defined as any set of organisms with a common ancestor.

Under PhyloCode, each clade's name would refer to a node in the tree of life and should thus provide nomenclature more appropriate for modern biological thinking, says the Smithsonian's Kevin de Queiroz, one of PhyloCode's developers. As such, it should simplify the current push to catalog millions of undescribed (and unnamed) species. "The inappropriateness and ineffectiveness of the current system in naming clades are obvious," asserts Philip Cantino, a plant systematist at Ohio University in Athens. "New clades are being discovered every day, but few are being named."

Defenders of the Linnaean system disagree, maintaining that its shortcomings—and the advantages of PhyloCode—are exaggerated. "PhyloCode is an impractical and poorly founded system," says Jerrold Davis, a systematist at Cornell University in Ithaca, New York. But Davis is worried nonetheless. "There's just one group of people standing on the street corner making a lot of noise," he says. Yet, "it's starting to consume resources and starting to appear in the popular press as if these folks have won."

Taxonomic Tradition

The Swedish botanist Carolus Linnaeus could not have anticipated the uproar that has erupted concerning the classification and nomenclature system he described in a 1758 book called *Systema Naturae*. At the time, names tended to be strings of descriptors that varied in length

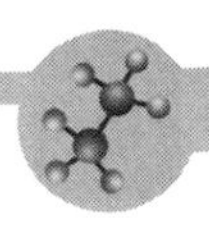

and meaning depending not just on the characteristics of the plant or animal but also on the scientist who named it. To enable botanists to equate plants from Europe, say, with plants from Turkey, Linnaeus devised a standardized system that has grown into the modern genus-species designation.

He also came up with basic principles for organizing newly named species into groups and then for assigning groups to specific taxonomic categories. His followers shaped this classification system into the "ranks" that have since been taught in every basic biology class. Thus humans are *Homo sapiens*, part of the genus *Homo*, the family Hominidae, the order Primate, the class Mammalia, the super-class Tetrapoda, the sub-phylum Vertebrata, the phylum Chordata, and the kingdom Animalia.

But in Linnaeus's mind, a species never changed—Darwin's observations about variation and evolution were still a century away. Thus, the Swede's system made no provision for naming and classifying organisms with evolutionary relationships in mind. "The Linnaean system was set up under a creationist world view to reflect a hierarchy of ideas in the eyes of the creator," explains Brent Mishler, herbarium director and systematist at the University of California, Berkeley. Furthermore, as far as Linnaeus could tell, life consisted of about 10,000 species. "The world was much more circumscribed than [the one] we have today," points out systematist Peter Stevens of the Missouri Botanical Garden in St. Louis.

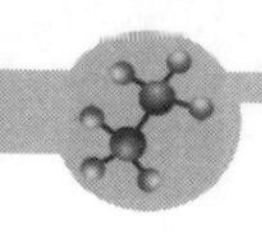

Darwin's 19th century contemporaries maintained the Linnaean system. But as they learned more about the number and evolution of organisms, they found they had to devise ever more extensive nomenclature rules, or "codes"—one each for plants, animals, and microbes—that would guide researchers as they fit new species into the traditional ranked hierarchies and enable them to keep names and classifications straight.

Under the traditional system, a taxonomist begins by assessing the physical characteristics—say, petal or leaf arrangements a set of species has in common—then selects the most representative species to be the "type" for each genus, then the most representative genus to be the type of the family, and so forth. Individual specimens are then deposited in a museum to serve as the reference point for that species and genus. Thereafter, as new specimens with similar characteristics are found, they are deemed part of a known species, a new species, or even a new genus based on how closely they resemble the type specimen. In this way the original "type" becomes an anchor point for the ranked groups to which it belongs. Thus, the flowering plant *Aster amellus* is the type species for the genus *Aster*, which in mm is the type for the family Asteraceae and the order Asterales.

Because of this dependence on type species, if a systematist reassesses a group of organisms and concludes that certain members don't belong, this

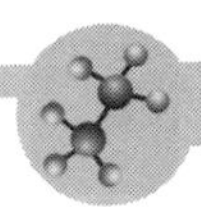

removal can sometimes mean that the group's name must change, or a new group name must be created. Thus, when a group of herbs called Ajuga was added to the family called Teurcrioidae, that family had to be renamed for the herb's original family, Ajugoideae, because it was the older name. And because a common weed called henbit (Lamiurn arnplexicaule) is the type genus for the mint family, any subfamily with Lamium in it must go by the name Lamiodeae. But over the past 35 years, Lamium has been reclassified three times—with each reclassification putting it into a new group. Thus three different subfamilies have borne the name Lamiodeae at one time or another. This ambiguity and these name changes are a hassle, say PhyloCode's advocates.

Another problem is that researchers unfamiliar with the intricacies of ranks often misinterpret them. For example, sometimes biodiversity is assessed in terms of numbers of families, but that ranking says little about the number of species contained therein. The family Hominidae has only one living species—*Homo sapiens*—while other families have tens, hundreds, or in the case of some plant families, even thousands of members. Because ranks are not always equivalent, a simple family count may give a false picture of an area's biodiversity. Even Peter Forey, a vertebrate paleontologist at the London Natural History Museum who supports the current naming system, agrees that "the Linnaean ranks . . . don't mean a lot in the modern-day world."

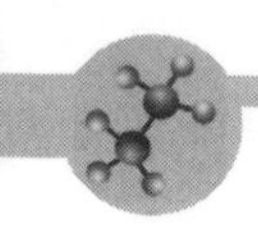

Time for a Change?

These drawbacks became apparent to de Queiroz in the 1980s while he was a graduate student at the University of California, Berkeley. At the time, a new way of classifying organisms called cladistics, based on assessing the evolutionary histories of features shared by organisms, had begun to make its mark on the field. This was causing great rifts among systematists about how they should do their work, as existing Linnaean categories were not based on phylogeny. Like increasing numbers of his contemporaries, de Queiroz wanted to reclassify the organisms he worked on following the principles of cladistics; yet he wasn't sure how to apply the existing nomenclature codes to the groups, or clades, he came up with.

As a result, de Queiroz says, applying the existing nomenclature codes could be cumbersome and confusing. As he and his colleague Jacques Gauthier, now a vertebrate paleontologist at Yale, were writing up their work, he recalls, "we stumbled on the idea of developing a naming system depicting phylogenetic relationships. At the time we didn't realize the full significance of it."

As de Queiroz and Gauthier worked out the conceptual underpinnings of such an approach over the next 8 years, they began to wonder whether Linnaean taxonomy had out-lived its usefulness. Thus was born the PhyloCode, and from the start, it didn't quite jibe with the Linnaean approach. For example, one way a PhyloCoder might define a clade would be to choose the two most distantly related organisms in that group as

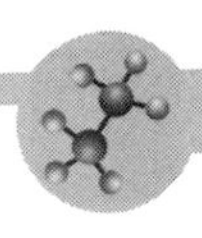

the "specifiers" and say that the group consisted of all those with the same last common ancestor as the specifiers. Such groupings didn't always coincide with previous membership in ranks.

The researchers described these ideas in several publications during the early 1990s, and then introduced them to the broader biological community at a symposium held during the 1995 meeting of the American Institute of Biological Sciences (AIBS). Interest was strong enough that de Queiroz and his converts organized a workshop at Harvard in August 1998.

Among the 30 attendees was Ohio's Cantino, who 4 years earlier had "written off phylogenetic nomenclature as impractical," he recalls. But for the AIBS symposium, he had been asked to evaluate how the old and new approaches would work with the mint plants that he studies. As a result, Cantino says, "I realized that phylogenetic nomenclature has great advantages."

Cantino has since become one of the PhyloCode's strongest advocates. He helped de Queiroz polish rules for the new system that were developed at the 1998 workshop. In May 2000 he posted them on the World Wide Web for comment (www.ohio.edu/phylocode). As comments trickle in, momentum is building to establish a society to guide PhyloCode's continued development, says Yale's Donoghue.

Tough Sell

Whereas almost all 21st century systematists now take a phylogenetic approach toward classifying

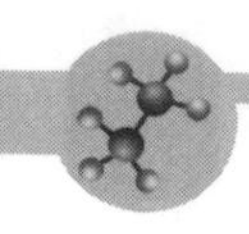

organisms, PhyloCode presents them with an alternative to the Linnaean approach for naming what they classify. One key difference is that because organisms would be grouped in clades under the new system, names would include no references to families, orders, classes, even genera in the traditional sense. And the definition of each name, be it for a species or some more inclusive clade, would be based on the shared ancestry of its members.

PhyloCode advocates haven't settled what will happen to species names, but they insist that most Linnaean family, class, or order names will survive the transition and will usually cover the same array of organisms. Thus, there could be a clade called Asterales that included another, smaller clade called Asteraceae, and the traditional relationship of these two groups would be retained. "Critics have said you'd lose all the hierarchical information, but you wouldn't," says Berkeley's Mishler.

In addition, PhyloCoders say that once a name has been redefined in PhyloCode terms, it should be more stable than it has been trader Linnaean rules. For example, in the PhyloCode system, the addition of the herb Ajuga to Teurcrioidae would not have forced that name to be changed. Unlike in the Linnaean system, they say, the new definitions will allow for organisms to move in and out of clades without disturbing the clade's name or the names of the other organisms. In some ways, "PhyloCode is a more flexible naming system," Missouri's Stevens asserts.

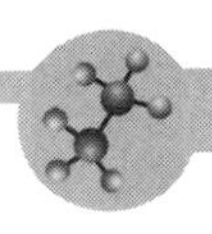

Both sides agree that the names of living organisms should be stable. "You don't want a system of nomenclature that is too mushy, where the names have no meaning," says the Smithsonian's Frank Ferrari, a PhyloCode critic. But both sides vehemently disagree about which system provides the strongest guarantee that a name and its meaning will remain unchanged through the decades. In the Linnaean world, instability arises because names for the groups change as the group's members change. Yet in the PhyloCode world, say its critics, names may stabilize, but what they signify will change as new evolutionary studies cause members to shift from clade to clade—as is bound to happen.

Evolutionary biologists across the globe are busy rearranging many branches of the tree of life, often by comparing genetic material from a wide range of species. Sometimes analyses of one gene will lead to a different branching pattern than analyses of a different gene. Organisms, perhaps even those specifying a clade, may shift in and out of the clade accordingly. Thus, PhyloCode "is not stable to changes in the phylogeny," Cornell plant systematist Kevin Nixon contends. And he argues that Linnaean categories may be just fine, as they are being revised to better reflect phylogeny.

Nixon and others also see value in retaining the Linnaean ranks, even if they lack biological meaning. "They are extremely important to our ability to communicate information about the biodiversity that we see and study," he argues. When he teaches a class, he likes to be able to refer to families, so that if he's taking a class on a field trip, he can communicate about whole groups of trees. Clunky as the current system may be, it works, he insists, because of what family, class, genera, and other names have come to represent.

Despite being convinced that the Linnaean way is superior, Nixon is concerned about the headway PhyloCode is making. "[PhyloCode] is not going to die out, because the spinmeisters behind this have the ear of the large funding agencies," he complains. Even the upcoming workshop on Linnaean taxonomy at the Smithsonian National Museum of Natural History "will very much play into the PhyloCode [camp's] hands," Nixon predicts. At any rate, there's likely to be vociferous debate about the two systems at the meeting.

But what rankles Nixon and his loyal Linnaean colleagues the most, they say, is that PhyloCoders appear to have seceded from the taxonomic community. Several governing bodies exist to help enforce and clarify Linnaean codes of nomenclature, but PhyloCoders seemed to have bypassed both the codes and their congresses. "They are going to erect a shadow government and [set up] a coup," Nixon complains. "This is arrogance."

In their defense, PhyloCode supporters say they have no choice but to go outside the existing system. "The differences between phylogenetic and rank-based nomenclature are just too fundamental for them to be combined," Cantino argues. Furthermore, they say they need an organization that can help iron out the details of PhyloCode. One contentious issue: how to name species.

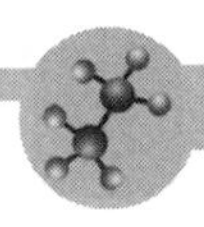

Many systematists, such as Stevens, want the names to remain the same. "The only reason to junk [a name] would be because it causes widespread confusion," he suggests. "You can add lots of higher order stuff by PhyloCode around the rudiments of the Linnaean system."

But in one popular proposal, just the "species" epithet would become the name. So *Homo sapiens* would get shortened to sapiens. Its drawback: Many organisms would need further qualification, as there are quite a few genera, for example, with a species named vulgaris, and searching archived literature for the "vulgaris" organism could yield many false citations. Another proposal calls for adding a number that might signify the place of a particular vulgaris on the tree of life, while a third calls for simply adding a hyphen to the existing genus-species designation, thereby linking them for all time. Although this yields a stable, unambiguous name, it could be misleading should phylogenetic studies later prove that one species didn't really share a common ancestor with another with the same genus name.

The lack of agreement about what to call species gives many systematists pause, even those who are open to PhyloCode. PhyloCode is "not ready for prime time," says Paul Berry, herbarium director at the University of Wisconsin, Madison. But the only way PhyloCode will make it to prime time will be if systematists take it seriously enough to test its potential. "One can never know for sure that one system is better than another until both have been tried for a while," says Cantino, who is

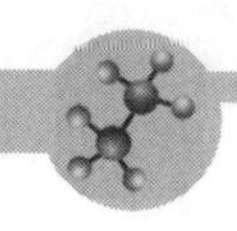

nonetheless pleased about the volume of activity at the PhyloCode Web site. "Most of the negative reactions have come from people who have not visited the Web site," he notes.

He anticipates that continued feedback will lead to refinements, and that over the next several years, researchers will start naming organisms using both approaches. In this way, the relative shortcomings and merits of each will become apparent.

Elizabeth Pennissi, "Linnaeus's Last Stand? (Introducing PhyloCode)," SCIENCE 291: 2304-7 (2001).

The Problems of Classifying Plants and Animals

2

Taxonomy helps scientists understand the relationships among different organisms. Initially, taxonomy was seen as a simple process of identifying new species and placing them in the correct position on a hierarchical tree of life based on their physical characteristics. Over time, however, it has become clear that the process of categorizing organisms is quite complex.

Good taxonomy can increase our understanding of the organisms around us. Poor taxonomy can have a serious effect, as was the case with the importation of gypsy moths under the mistaken impression that they were a type of silkworm. Ultimately, this led to the destruction of a vast number of trees.

The following article discusses one of the latest techniques—"cladistics"—for classifying organisms. The cladistic approach uses everything that is known about an organism—its biology, behavior, molecular profiles, biochemistry, and fossil history

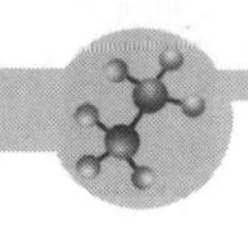

(if this information is available)—to create a mathematical diagram called a cladogram. A cladogram shows how similar two organisms are. Such cladograms can reveal surprising relationships between organisms. —JF

"Who's Your Daddy? The great apes are among the most popular animals in most zoos. Their actions, facial expressions, and family life remind us so much of ourselves. Have you ever wondered, though, how we might look to them?"
by Stephen Whitt
***Odyssey*, December 2003**

A modern, statistically based system of classification called cladistics reveals a surprising possibility: Not only are humans apes, but we are more closely related to some apes—namely the chimpanzees—than the chimpanzees are to orangutans or gorillas. In other words, the chimpanzees' closest living relatives are—us! A gorilla, looking past our culture and technology, might see us humans as just another type of chimpanzee.

What Is Cladistics?

Cladistics is a rigorous, mathematical representation of a simple idea—that all living things are related. These relationships are expressed in diagrams called cladograms.

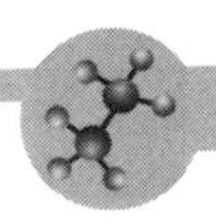

The simplest cladogram links two living things—you and a chimpanzee, for instance. The diagram doesn't seem to display much information, because there's only one way to draw it.

You and a chimpanzee have a lot in common—sharing among other traits four limbs, hair, a big brain, and 99 percent of your DNA.

In other ways, you and a chimpanzee are different—that's why you appear on separate branches of the cladogram, after all!

The interesting questions arise when we add a third living thing to the cladogram—let's say a gorilla. There are three different ways to draw this relationship. How do we decide among them?

The answer is by comparing lots of traits, and then seeing what the statistics say. Some traits, such as hair, are shared by all three of you, so they're of no help. Other traits, such as knuckle-walking, are shared by only the chimpanzee and the gorilla. Still others, including the greatest degree of DNA similarity, belong to you and the chimp.

By comparing many traits, and by analyzing the results with sophisticated statistical measures, researchers can determine which of all the possible cladograms is the simplest—or in the language of cladistics, the most parsimonious. ("Parsimonious" comes from the Latin verb *parsere*, "to spare.")

One of those statistical measures is called the consistency index (CI). Another is called the retention index (RI). Both rate the cladogram's simplicity

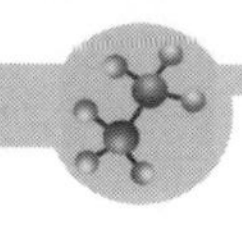

on a scale from zero to one. The simpler the cladogram, the closer to one its CI and RI will be. Why this interest in simplicity? According to the logic of cladistics, the simplest cladogram is the one most likely to be "true."

So what does the data reveal? The simplest (and therefore the most likely) answer to the human-chimp-gorilla question is that humans and chimps are close cousins—closer to each other than either is to the gorilla.

Surprise, Surprise

The application of cladistics has led to some other big surprises:

- Almost all scientists today accept the idea that birds are actually dinosaurs. Cladistic analysis of living birds and extinct meat-eating dinosaurs called theropods shows a close connection. The power of cladistics has convinced nearly all the experts that the dinosaurs are with us still.
- Horseshoe crabs are ancient-looking sea creatures that are often called "living fossils." They look something like other shelly ocean dwellers such as lobsters and crabs. But cladistic analysis shows that horseshoe crabs are most closely related to—are you ready for this?—spiders! The idea that a lumbering creature from the sea floor is cousin to those nimble web-spinners has to

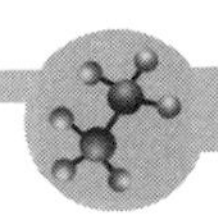

be one of the most surprising results of cladistics.

- Finally, humans and other vertebrates (animals with backbones) form a related group. But which of the invertebrates—groups such as insects, mollusks, and worms—is our closest relative? This debate raged for years. Today, cladistics reveals that animals called echinoderms, such as the slow-and-steady starfish, are our own long-lost invertebrate cousins. Without cladistics, we never would have guessed.

Fish Tales

In some cases, cladistics turns common knowledge on its head. For instance, we all know what a fish is, right? Well, maybe not. Cladistics reveals that fish with fleshy lobe fins (instead of the ray fins most fish have) are our close relatives. At some time in the past, one species of lobe-finned fish branched off from the rest in an event that led eventually to us. But at what point did our ancestors stop being fish?

The answer from cladistics is: never. Once a fish, always a fish—albeit a very peculiar air-breathing four-limbed, bicycle-riding fish!

Cladistics shows that not only are we apes, but we and other mammals, as well as reptiles, birds, and amphibians, are all highly modified, lobe-finned fish. Think about that the next time you take a dip in the pool!

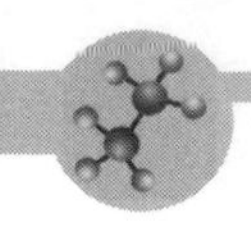

What Is "True" in Science?

What is "true" in science? Scientific knowledge is never certain. Scientists know they have an incomplete picture of the world. Scientists see the past through the evidence of the present, and that evidence is always imperfect.

It is possible, for instance, that the shared traits between humans and chimps are coincidental, and don't truly reflect a recent common ancestor. Perhaps humans and chimps evolved those traits separately (this is called convergent evolution).

As more and more data is gathered, however, the close relationship between humans and chimps becomes ever more certain. Yet no matter how much data scientists collect, they will never be completely sure they have the right answer. —S.W.

Jonathan Coddington, the chairman of the entomology department and curator of spiders at the Smithsonian's National Museum of Natural History in Washington, D.C., is a specialist in spider behavior and taxonomy. Using Coddington and the Smithsonian's collection of spiders and other entomological specimens as examples, Sue Hubbell, the author of the following article, explores the types of questions that taxonomists face today. Hubbell explains how taxonomy is

being used by scientists to increase our knowledge of the natural world. —JF

From "How Taxonomy Helps Us Make Sense Out of the Natural World." (Spider Collection at the Smithsonian's National Museum of Natural History)
by Sue Hubbell
***Smithsonian*, May 1996**

In the past months I've come to know, tolerably well, a big, beautiful spider. She is blotchily orange and tan with darkly banded legs. Each day, she spins a fine new round web somewhere in the garage. Her eyesight is none too good and she usually sits off the web, hidden against a protecting beam, but when moths and flies blunder into her trap, she can feel the vibration on one of the web's guying threads and she rushes out. She eats the first and wraps the others in silken winding sheets to keep for later. I try to avoid tearing her web and save her repair work, but I know she is a quick and efficient spinner. Jonathan Coddington, the chairman of the entomology department and curator of spiders at the Smithsonian's National Museum of Natural History, tells me she eats each day's web and reprocesses the protein. Within 20 minutes of munching it down, she can spin recycled silk.

Her common name is barn spider. Her scientific one is *Araneus cavaticus*. Her genus, indicated by the first word in her scientific name, includes a greater number of species—more than 1,500—than any other spider genus. The genus name and that of her order,

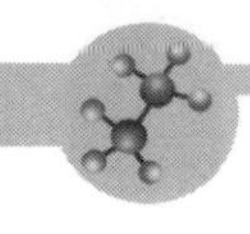

Araneae (the spiders), as well as that of her class, Arachnida (spiders plus a lot of kinfolk: ticks, daddy longlegs, scorpions and such like), echo the name Arachne, borne by a Lydian princess who was such a skilled weaver that Athena grew jealous of her. Terrified of the goddess wrath, Arachne hung herself from a rafter, and Athena transformed her into a spider and her rope into a web.

. . . Back at the beginning of time, when I took introductory biology, living things were divided into a neat hierarchical series of inter-nesting boxes, called taxa, with the higher taxa containing the lower: kingdom, phylum, class, order, family, genus, species. Of the first, kingdom, there were two: animal and vegetable, although even then bacteria were something of a problem, off to one side. The whole lot was presented in an ascending order with mankind triumphant at the top. In those days, taxonomy was mainly a matter of identifying and naming what few species remained to be discovered.

Well, the world hasn't changed, but our understanding of it has, and taxonomy is no longer so simple . . .

It turns out also that there are more species we don't know anything about than we ever dreamed of in introductory biology. Vertebrates, of which we are one kind, are big and obvious and fairly well known, but better sampling techniques are finding millions of new invertebrates from the treetops to the ocean bottoms.

Taxonomists of Jon's sort ask questions about what is going on at the higher taxonomic levels: What constitutes

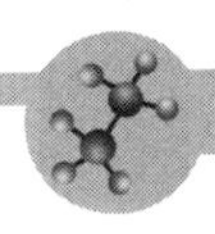

spiderishness? How does one family of spiders differ from another? What is the evolutionary relationship of taxa of spiders that spin different kinds of webs? Asking questions like these was the way Jon overturned a long-held notion about spider phylogeny, or evolutionary history.

Spiders spin a variety of webs peculiar to their own kind. Sheet-web spiders, for instance, make webs that look like flattened hammocks. Theridiid spiders are the ones that spin those cobwebs that tidy people dust out of basement corners. My *A. cavaticus* and others spin the familiar orb-shaped webs.

It had long been assumed that orb webs were the highest achievement of spiderly craft and were, adaptively speaking, the most advanced. Such webs were thought to be maximally efficient, which is why their production had evolved independently in different taxa of spiders . . .

After careful observation, however, Jon inferred that rather than being the ultimate web, the orb web is primitive, ancestral. The cobweb is actually much denser, more protective and more efficient for trapping prey than the orb web, and more elaborately engineered, as well. He concluded that the cobweb evolved from the orb web, rather than vice versa. The earlier spider classifications that separated orb weavers into differing taxa based on anatomical differences have been generally scrapped, Jon tells me, because consolidating all orb weavers in a single line is a much better representation of what we know about their behavior . . .

I am sitting in his office on the third floor of the Natural History museum as he explains this to me. The

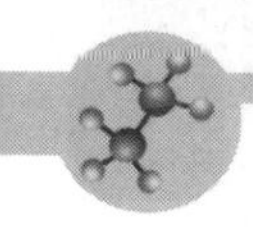

sun floods through his windows overlooking the Mall beyond . . . Every year more than five million people visit the museum. They remember the elephant in the rotunda, the nice lady at the information booth or dinnertime for the tarantula in the Insect Zoo. But what they do not see is the undisplayed 99 percent of the collections, numbering 122 million objects, or the 400 scientists and technicians who with them. To be precise, there are 108 Smithsonian scientists and another 25 to 30 scientists from federal agencies whose research duties keep them in the museum full time. Together, the scientists, support staff and collections make up the largest research museum in the nation and one of the preeminent in the world.

Jon's sunny office is filled with books, files of arachnological papers and a scattering of vials containing spiders afloat in alcohol. He shows me a flat container with a real web in it—that of a brown recluse spider, whose bite is as dreaded by many people as that of a black widow. "Good taxonomy," says Jon, "has predictive value, which is often useful. For instance, there is a spider in South Africa known as *Sicarius*—the name means 'murderer'—that can give a serious bite. No one has done much work on it, but we do know it is in the family Sicariidae, which is, taxonomically, the sister group of the family Loxoscelidae, the family to which the brown recluse belongs. We do know rather a lot about the brown recluse, so we can make predictions about the biology of *Sicarius* and know how to treat the bites."

Bad taxonomy, on the other hand, can be expensive. Many millions of dollars have been spent to eradicate the

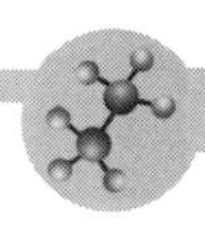

gypsy moth in this country; it was imported from Europe a little more than one hundred years ago as a consequence of mistaken taxonomy. Leopold Trouvelot, a French amateur naturalist and astronomer, was using the gypsy moth in an attempt to develop a better silkworm when it escaped from his laboratory near Boston. Undeterred by natural predators, it has been eating its way through our Eastern forests ever since. In Trouvelot's time, the gypsy moth was classified in the genus *Bombyx*, that of the silkworm, which was and still is *Bombyx mori*. The gypsy moth was, but no longer is, *Bombyx dispar* (meaning a silkworm with males and females of different color). Trouvelot probably would not have experimented with it had he known it by today's name, *Lymantria dispar*. *Lymantria* means "destroyer."

On Jon's desk are specimen spiders that have been collected in the Cameroons. Jon tries to spend about a third of his time in the field collecting spiders and studying their behavior. He passes the spiders, along with the exact location where they were collected, to a colleague who enters information about them into the computer record and prints out labels that identify them and tell the who, when and where of their collection. Groups of individual vials containing the new specimens are packed into straight-sided, half-liter, alcohol-filled bottles that resemble old-fashioned home-canning jars because they are closed with a rubber gasket and a metal clamp.

Jon takes me back to a room filled with tan metal cabinets that look like map cases with fat drawers. Within the drawers the half-liter jars are stored,

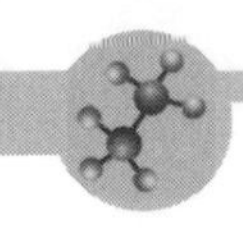

arranged alphabetically by family and subarranged by genus and species, if known, and geographical origin. He pulls out a drawer containing my *A. cavaticus*. There are two vial-filled jars of them, but many more of other species of Araneus. All told, Jon estimates, the Museum's spider collection contains some 116,000 specimens.

. . . Specimens, preserved and classified, make up what is called a "synoptic" collection. That means that it is a synopsis, a holding of representatives of major taxa from all over the world. "This collection," says Jon, "contains one of the greatest synoptic collections of insects and their relatives on Earth. This is where diversity can be studied."

. . . It was not until the 1960s, in fact, that evolutionary, or phylogenetic, classification came into its own. The system that eventually emerged is known as cladistics, and while not accepted 100 percent by taxonomists, it is now in use at major museums and universities throughout the world.

Cladistics is a "laser-like effort" to find the one true tree, Jon says. It defines a given taxon by making use of unique traits, believed to be of such recent evolutionary origin that they set the members of that taxon apart from any other taxa on the family tree. These traits are drawn from summaries of everything known about the animals within the taxon: their biology, behavior, molecular profiles, biochemistry, and fossil history if known. When the animals defined by specific traits are grouped, therefore, it is assumed that they diverged from a common ancestor and are closely related.

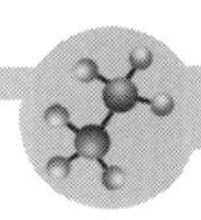

"Cladistics," Jon says, "is the quantitative analysis of comparative data used to reconstruct evolutionary trees. It took numerical analysis and wedded it to our biological understanding of evolutionary history. It is also simply the most efficient way that has yet been devised to store and retrieve biological information."

On a simple level, here's how it works. Let's say that when you are studying the order of spider, "eight-leggishness" would not be considered a "derived" trait because all of the class Arachnida—ticks, daddy long-legs and mites, as well as spiders—have eight legs. For this purpose, it is a primitive trait that emerged early in the evolutionary history of the class. You would define spiders as an order by other, more recent characteristics, two of which might be "having silk-spinning organs at the nether end of the body" and "having fangs equipped with poison glands." For cladistic analysis within the order of spiders, those two traits would become primitive because all spiders possess them, and other characteristics would be looked for to sort out families, genera and species.

I asked Jon to give me an example of what separated a cobweb spider from my barn spider, *A. cavaticus*. He told me that cobweb spiders (and a few of their relatives) hurl huge droplets of sticky silk to attack their prey. *A. cavaticus* can only hang sticky silk in its webs. The sticky-silk attack is always the same in cobweb weavers. It is unique. It is a derived trait.

Jon pulls from under a pile of papers a computer printout that he calls, proudly, "*the* matrix," which contains 49,000 observations of spiders accumulated to date by him and his colleagues. It applies 354 traits or characteristics along one axis to 139 genera of spiders along the other axis. Points in the matrix include spiders' body structure, behavior, biochemical makeup, web-building patterns—anything that can be found in the literature. These 49,000 observations are coded digitally so that comparisons may be sorted out by computer. The family trees worked out from these comparisons, called cladograms, are visual representations of this analysis. Having them enables us to make major predictions about newly discovered organisms, especially "exotics," those that have arrived from elsewhere. The cladograms tell us instantly what other organisms share the same behavior. In agriculture and disease control, such knowledge can be worth billions.

The origins of cladistic analysis lie in the work of a little-known German entomologist, Willi Hennig, who died in 1976. Hennig wanted to create a taxonomic system that emphasized phylogeny (the evolutionary history of a genetically related group of organisms) in groupings by using recent characteristics instead of older ones.

During World War II, Hennig was held prisoner by the British in Italy. While he was there he wrote a book in which he sketched a "phylogenetic systematics" that, he hoped, would prove to be nothing less than "the general reference system of biology."

The book was largely unnoticed until the late 1960s, when a revised edition was translated into English as Phylogenetic Systematics. A group from the American Museum of Natural History adapted the

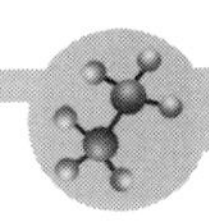

computer techniques of the numerical taxonomists to Hennig's ideas. Numerical taxonomy is a kind of computer systematics that uses all of an animal's characteristics to create an "objective, repeatable classification scheme."

As taxonomists continue to identify more and more new species and revise ever upward their estimates about how many others remain to be discovered, taxonomy itself has come increasingly to resemble a numbers game. In Coddington's specialty alone, there are today something better than 36,000 known species of spiders and perhaps another 100,000 as yet unknown.

"Look," Jon says, "there are only 7,000 families of all kinds of life on this planet. Of those 7,000 families, there are 105 that are spiders. We have, right here in the museum, representatives of nearly all of them." Taking it to the next level, he tells me there are 3,000 genera of spiders of which the museum has about 600 named and identified (many more are waiting in those jars). He is working fast to collect the rest of them before they are extinguished, to fill out the synopsis the world's museums hold. Jon is an engaging man who wears his erudition lightly. He smiles easily, but when he speaks of this aspect of his work as curator, he is intense, serious.

"You know what I want to do?" Jon continues. "I want to create, as quickly as possible, a synoptic collection of the chunk of life for which I am responsible. Because 200 years from now there's going to be a news conference held here, right in the museum. There will be the Minister of Environmental Affairs, who will

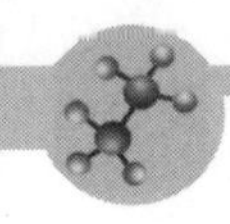

announce that the inventory of life on Earth is complete: 5,748,941 species. And some reporter will stand up and ask, 'But Madame Minister, didn't they say 200 years ago that there were 12 million, 15 million, 30 million, even 80 million species?'"

"Madame Minister will be able to answer in only one of two ways. Either she'll say, 'Well, we lunched them all,' or she'll say, 'They didn't know what they were doing 200 years ago.' I want to make it impossible for her to give that last answer."

DNA Analysis and Taxonomy

3

One of the most exciting scientific advances of the late-twentieth and early twenty-first centuries has been the development of DNA sequencing technology. The use of genetic analysis techniques has allowed us to identity both present and evolutionary links among organisms in a way that was never possible with mere observation of physical characteristics.

The following article, "What's in a Name?" discusses the potential to use a single genetic sequence to uniquely identify organisms. This method is similar to the way bar codes are used to identify products in a store.

Dr. Paul Hebert and his colleagues at the University of Guelph, in Ontario, Canada, believe that if a system similar to the universal product code (UPC) can be developed, it would greatly speed up the process of identifying organisms. Plants and animals could then be quickly categorized by reading a small fragment of their DNA. Dr. Hebert feels that the use of

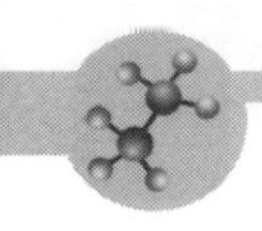

such technology would make it possible to inventory all known organisms and their respective codes within twenty years. —JF

"What's in a Name? Taxonomy. (Categorising New Species)"

by Staff Writer
***The Economist (US)*, January 4, 2003**

It might be better to identify species by number, not name. For thousands of years, humanity has classified the living things of this world in much the same fashion: by their appearance. If it looks like a duck, walks and quacks like a duck, then it is a duck. But tackling millions of species in this way has proven to be a recipe for confusion. As taxonomists have found at their cost, what looks like a duck may in fact be a goose.

More recently, genetic techniques have been applied, particularly for distinguishing the more difficult-to-identify species such as viruses and bacteria by comparing pieces of DNA. Might this approach be more generally applicable? Paul Hebert and his colleagues at the University of Guelph, in Canada, think it might be. Just as barcodes and the "universal product code" numbering scheme uniquely identify different items at a supermarket checkout, they suggest that some stretches of DNA could perform a similar function in living things. In a paper just published in Proceedings of the Royal Society B, they discuss how long such a genetic barcode needs to be, and where it might be found.

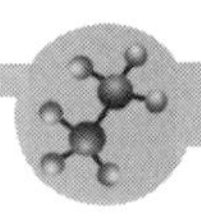

A universal product code found on the high street consists of a string of 11 digits, each of which is one of ten numerals, providing 100 billion unique combinations. Genetic material, however, uses a quaternary, rather than a denary, coding system. Every organism's genome is encoded using a quartet of chemical bases—adenine, cytosine, guanine and thymine, generally referred to by their initial letters, A, C, G and T—in a DNA sequence that can be millions of letters long. In theory, it would only be necessary to sample 15 of those letters to create one billion unique codes.

In practice, however, the characteristics of DNA mean that 15 letters are not enough. Unlike the arbitrary numbers of a universal product code, the letters of DNA are not random, because they code for something that has a biological meaning. So the researchers estimate that a 45-letter signature would be required. As luck would have it, determining the sequence of several hundred letters now costs no more than sequencing a few dozen. As a result, the researchers are confident that it will be possible to capture enough information to distinguish tens of millions of species, using existing technology.

But where is the best place to find a universal product code for organisms? Not within the genome inside the nucleus of living cells, surprisingly. Instead, the researchers suggest targeting the smaller genome found inside cellular components called mitochondria. Such mitochondrial DNA has several features that make it suitable for use as a genetic barcode. It is generally passed unchanged from parent to offspring, unlike

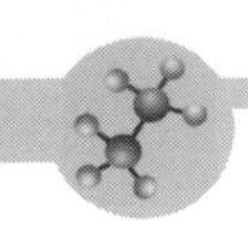

nuclear DNA in which maternal and paternal contributions are mixed and shuffled with each generation. It is also relatively free of long "non-coding" regions (sometimes called "junk DNA") that can cause confusion when comparing DNA sequences.

Within mitochondrial DNA, the researchers believe there are a number of possible genes that might be suitable for use as a biological universal product code. However, one gene in particular has caught their attention. It is called cytochrome c oxidase I and it plays a key role in cellular energy production. It is easy to isolate. Variations in its genetic sequence should work as a unique code to enable different species to be distinguished. Better still, comparison of different organisms' unique codes should help to show how different species are related, and how and when new species evolved.

Until a few years ago it was an immense task to get a useful DNA sequence from a specimen. Today, it is possible to find, cut and copy sequences so fast that you can go from the leg of a beetle to a mitochondrial DNA sequence in only a few hours. This will only improve in future with further automation; the use of dedicated DNA-chip arrays would speed things up even more. Assuming that a system akin to the universal product code can be devised and agreed upon, Dr Hebert says it should be possible to compile a complete inventory of known organisms and their corresponding codes within 20 years.

This would revolutionise taxonomy, which began 250 years ago with Linnaeus and has so far managed to categorise only 10% of the earth's estimated 10m-15m species. As well as providing a short cut to the taxonomic

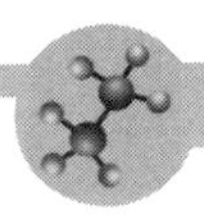

finishing line, an inventory of genetic barcodes would dramatically simplify and speed up the process of identifying organisms from small samples. Ultimately any person, with only an afternoon's training, would be able to identify an organism from just a small fragment. The ability to read nature's barcodes could have as much of an impact in the laboratory as man-made barcodes have already had in the shops.

This article, despite its amusing name, provides insight into how scientists are using certain types of DNA to trace the evolutionary ancestry of animals. Gregory McDonald, a paleontologist with the National Park Service in Denver, Colorado, and his colleagues have managed to extract DNA from the fossilized dung of large ground sloths found in a cave in southern Nevada. The article also illustrates how such genetic analysis techniques are leading scientists to change their ideas of the proper classification of specific animals.

For example, in the past, scientists have had trouble establishing the exact relationships between living and extinct sloths based solely on comparisons of skeletal characteristics and mitochondrial DNA. The characteristics of the newly analyzed nuclear DNA are helping to clarify issues such as whether two- and three-toed

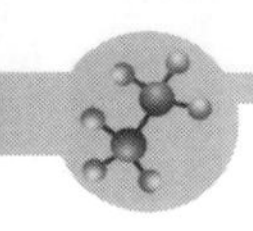

sloths both evolved from tree-living ancestors or whether they evolved from ancestors with distinctly different living patterns. —JF

"Ancient Poop Yields Nuclear DNA (Secrets of Dung)"

by Sid Perkins
***Science News*, July 12, 2003**

Researchers have extracted remnants of DNA from an unlikely source: the desiccated dung of an extinct ground sloth that lived in Nevada at the height of the last ice age. The feat is the first recovery of genetic material from cell nuclei of fossils that haven't been sheathed in permafrost. It suggests that scientists may be overlooking caches of fossil DNA preserved in warm arid environments.

Earlier work on fossils had isolated DNA carded in mitochondria, the powerhouses of living cells. However, the DNA in a cell's nucleus is typically longer and therefore holds much more genetic information about the species and individual from which the cell derived, says Gregory McDonald, a paleontologist with the National Park Service in Denver.

Now, McDonald and his colleagues have isolated snippets of nuclear DNA from a coprolite—or piece of fossilized dung—of a Shasta ground sloth, a 2.3-meter-long, 350-kilogram herbivore. The coprolite, found in a cave in southern Nevada, may be as much as 15,000 years old, says McDonald. The team's analyses suggest that as many as 4,000 fragments of nuclear DNA

measuring at least 100 base pairs in length may be present in each gram of the animal's desiccated feces. The researchers report their findings in the July 1 Current Biology.

Scientists using only skeletal characteristics and comparisons of mitochondrial DNA have had trouble discerning the relationships among extinct and living sloths, says McDonald. Some of those studies have suggested that the tree-dwelling lifestyle of all living varieties of two-toed and three-toed sloths evolved only once. Other findings suggest that arboreal living arose separately in two-toed and three-toed sloths.

The newly analyzed differences in nuclear DNA suggest that the Shasta ground sloth is more closely related to living three-toed sloths than to the two-toed varieties, which lends credence to multiple origins of tree dwelling.

The key to the preservation of the Nevada sloth's nuclear DNA was aridity, says McDonald. Lack of humidity desiccated the dung and stymied bacterial degradation of the genetic material in the stable environment provided by the surrounding cave.

The mummifying environment seems to promote long-term preservation of DNA despite the warm conditions, says Julio L. Betancourt, a paleoecologist with the U.S. Geological Survey in Tucson.

In separate analyses of ancient sloth dung from a sheltered ledge outside a cave in the arid foothills of the Argentine Andes, Betancourt and his colleagues may have identified a previously unknown species of extinct ground sloth. Mitochondrial gene sequences extracted

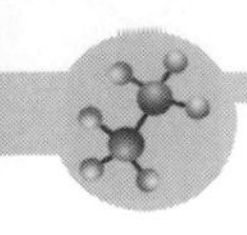

from cells in the 16,000-year-old, pecan-size pellets don't match those garnered from the four other living or extinct sloth species that have been genetically sequenced to date.

The Argentine dung fragments are much smaller than those left by the extinct horse- to elephant-size ground sloths already known to have inhabited the region, Betancourt notes. The dung came either from a species of ground sloth for which bodily remains haven't yet been found or from a species that also lived in another area but that scientists haven't yet genetically sequenced. Betancourt and his colleagues reported their findings in the *May Quaternary Research*.

This article, "'Deep Gene' and 'Deep Time,'" discusses how scientists are using genetic techniques to create a picture of the evolutionary history of plant species. Plant biologists have undertaken new initiatives aimed at developing a better understanding of plant evolution. By using gene-sequencing technology, scientists have been able to compare DNA sequences, such as those encoding ribosomal RNA and chloroplast proteins. This has resulted in a new understanding of the evolutionary relationships among plant species, which lead to a rearrangement of plants on the basis of their genetic relationships. Such

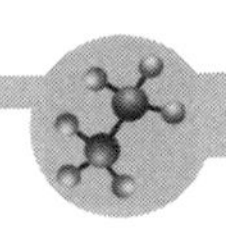

genetic studies have also allowed scientists to locate the base members of plant groups—the plants that were the ancestors of all other plants of a given type, such as flowering shrubs.

In one project called Deep Gene, plant taxonomists and molecular biologists are working with the genomes of plants such as rice. They are attempting to trace groups of genes that control specific processes such as flower development. Through this process, they hope to gain an understanding of the mechanisms that control major evolutionary changes. This article provides an excellent overview of the way that taxonomists are using genetic techniques to revolutionize the classification of plants. —JF

"'Deep Gene' and 'Deep Time'"
by Barry A. Palevitz
***The Scientist*, March 5, 2001**

Evolving collaborations parse the plant family tree.

Amid last month's hoopla over the human genome sequence and what it says about humans, plant biologists announced two new efforts aimed at a firmer understanding of plant evolution—who is related to whom and how—a discipline better known as systematics. Constructing evolutionary family trees is harder than investigating personal genealogies—biologists don't have the equivalent of birth registrations or family bibles to consult. Fossils tell them what ancient

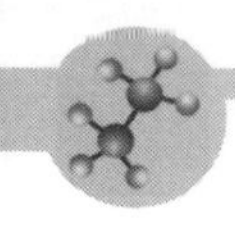

plants use to look like, but placing them in context with living organisms is difficult at best. Even the systematics of existing plants can be contentious, as researchers disagree on lumping plants together or splitting them apart in search of the most natural taxonomy.

Scientists liken constructing phylogenetic trees to tracing all the branches and trunks of a real tree, like an oak, with only characteristics of its outermost twigs to go on. That's because present day organisms are the sole survivors—called "terminals" by systematists—of multiple, diverging lineages. However daunting the process, researchers have made breathtaking progress in the last 20 years, thanks to gene sequencing. According to University of Georgia systematist David Giannasi, "it was a case of technology catching up with theory." By comparing DNA sequences such as those encoding ribosomal RNA and chloroplast proteins, systematists redrew large chunks of the plant taxonomic map.

A good example of the redefining process is found in the milkweeds, which taxonomists traditionally placed in a family called the Asclepiadaceae. They also thought the milkweeds were allied with a second family, the Apocynaceae. But based on molecular data, "the Asclepiadaceae nests within the Apocynaceae," says Giannasi, "so we now know they should be lumped together." The same is true for the mints, thought to be in their own family just a few years ago but now grouped with the verbenas.

Researchers have also clarified some of the most basal groups in the plant family tree. They now know that a previously obscure New Caledonian shrub

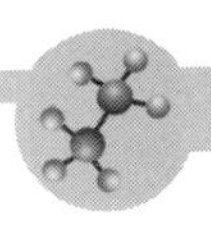

called Amborella is sister to all other flowering plants, or Angiosperms, with water lilies branching off the evolutionary trunk at the same level or just above.[1] They also think the gnetales, previously considered flowering plant allies, are probably more closely related to pines, in the Gymnosperms.[2] And horsetails and whisk ferns, once thought to relic descendents of early land plants, now seem more closely tied to the tree ferns.[3]

Feds Fertilize Interactions

One of the key ingredients in systematists' recipe for success was cooperation and communication. Thanks to joint funding starting in 1994 from the U.S. Department of Agriculture, Department of Energy, and National Science Foundation, a consortium of researchers called the Green Plant Phylogeny Research Coordination Group, or Deep Green, pooled ideas and resources in a joint plan of attack. Machi Dilworth, head of NSF's Division of Biological Infrastructure, thinks, "Deep Green was one of the very visible success stories" of the three agency effort. "With a little support they were able to come together and accomplish major scientific achievements."

NSF was so impressed with the collaborative approach, it decided to fund "Research Coordination Networks" (RCNs) serving all areas of the biological sciences. Like Deep Green, the grants foster communication and collaboration between scientists, but don't directly cover research costs funded by other programs. Two of the RCNs are scions of Deep Green.

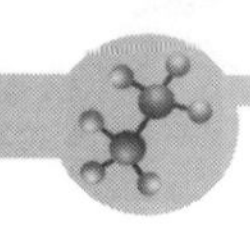

Systematists Dip into Genomics

In one of the team projects, called Deep Gene, systematists join forces with molecular biologists working on entire genomes like those of Arabidopsis and rice.[4,5] By tracing suites of genes that govern processes such as flower development, they hope to clarify mechanisms governing major evolutionary changes, including new biochemical pathways and the appearance of complex morphological characters. Sequencing also uncovers large-scale genomic changes including chromosomal rearrangements, which can be invaluable in defining plant relationships. Likewise, evolution depends on alteration in spatial and temporal controls governing gene activity—when and where genes turn on and off. The new RCN hopes to discover how gene regulation changed in the evolution of various plant groups.

Tolerance toward desiccation is a good example of how traits may have appeared and disappeared during evolution. The first plants to occupy dry land faced a big problem compared to their aquatic ancestors: an uncertain supply of water. Mosses, for example, grow in moist environments but also suffer periodic drying. That's why they require biochemical mechanisms that allow them to survive dry periods. When larger vascular plants arose, with roots and a plumbing system to extract water from the soil and move it long distances, desiccation tolerance became less important. But it reappeared later on in seed plants, which remove

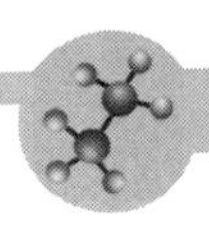

water from tissues surrounding young embryos in preparation for dormancy.

According to Deep Gene principal investigator Brent Mishler of the University of California at Berkeley—and a veteran of Deep Green—"around 80 genes are involved in desiccation tolerance in mosses. When desiccation re-evolved in seeds, some of these genes were reused." Mishler would like to know how such changes in gene regulation arose during major evolutionary events. Mishler chaired a symposium on Deep Green at the annual meeting of the American Association for the Advancement of Science, February 15-20, in San Francisco.

Daphne Preuss, molecular biologist at the University of Chicago and Deep Gene co-PI, says she brings to the table "the tools and techniques of high throughput, big scale biology." Still, in a true collaboration everybody benefits. With Deep Gene, genomicists like Preuss want to advance their own projects. In her case, that means figuring out how centromeres work. Centromeres are DNA sequences located where chromosomes attach to spindle fibers during mitosis and meiosis. Preuss has dissected centromeric DNA in Arabidopsis but knows that "the sequences are very diverse from organism to organism." The question is, "how did these differences evolve, and what key components are important for centromere function?" Adds Preuss, "I want insight from looking at conservation through evolution."

Preuss admits that "this is expensive work, so every decision counts. We're now making key decisions as to

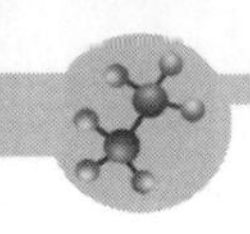

which species to look at next. We're looking to people in phylogenetics to help." Mishler sees other practical benefits from Deep Gene. "Can we use the information for agriculturally important plants that aren't desiccation tolerant?" he asks. By guiding researchers to promising sources, Deep Gene can also "predict useful chemicals for pharmacology," says Mishler. That makes University of Georgia's Giannasi smile because older studies comparing the chemical composition of plants—including substances such as terpenoids—predicted changes cemented by more recent gene sequencing projects. "The secondary chemistry was there, but nobody trusted it," comments Giannasi.

Fossils and Morphology Join the Fray

Doug and Pamela Soltis of Washington State University in Pullman lead another RCN called "Deep Time." Having done much of the gene sequencing for Deep Gene, the Soltis' want to superimpose other kinds of information on their phylogenetic trees, and in the process add the dimension of time to key points in plant evolution.

Years before systematists accessed gene sequences, they relied on other information in the form of morphological, anatomical and chemical characters. While valuable, such characters can be misleading. For example, a structural trait shared by two groups could have arisen by convergent evolution rather than common ancestry (though the same applies to DNA sequences). Moreover, the number of structural

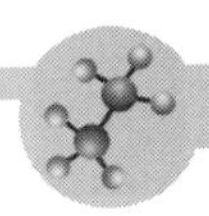

characters applicable to phylogenetic analysis is limited; DNA sequences, on the other hand, are far more useful since the average protein encoding sequence contains 1,000-2,000 characters, or nucleotides. That's why they turned to genes.

But the tide may be changing again, at least a little. The Deep Time RCN will arrange plants according to a "morphological matrix" of characters, but "constrain the taxa to conform to the DNA-based topology already available, and in which we have good confidence at this point," say Pam and Doug Soltis. They'll then "conduct a phylogenetic analysis of the morphological matrix with fossils included." The trick will be to pick characters from existing plants that also apply to fossils. Despite the fact that "fossils have rarely been integrated in a phylogenetic context for any group," the Soltis' are hopeful. Since dates are available for many of the fossils, their inclusion adds a time factor to the phylogenetic tree—systematists can assign dates to key branch points. They'll also integrate data from molecular clocks governed by mutations. "It's sort of like the movie Back to the Future," note the Soltis', "Having the timing of a key event in the past nailed down is critical in understanding what has occurred to produce what we see in the present."

The Soltis' also wax philosophical about the collaboration: "We spent a decade in the area of systematics largely focused on molecules. There is a wealth of information in nonDNA characters such as morphology and anatomy, and we can't lose expertise in these areas."

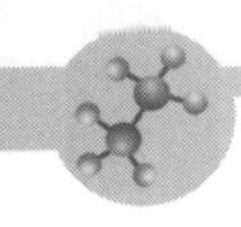

Problems? Cooperation Is the Key

Deep Gene and Deep Time researchers realize that reaching their goals may not be easy. According to the Soltis', "two big issues are missing data and the combinability of molecular and morphological data sets." Mishler agrees: "We don't know entirely how to do it. Theory hasn't kept pace—it's dealt mostly with sequence data." Researchers hope the latest collaborations will foster development of new methods to tackle such problems. Mishler sees promise. "The RCN will help us. Even a small amount of data from these other sources can improve phylogenetic trees" and eventually "lead to more research funding." The depth of cooperation is all the more impressive because Deep Gene and Deep Time will interact.

The "Deep" projects testify to the importance of collaboration in modern research. According to Doug Soltis, "the cooperative nature of botanists has really turned the tide in the past decade." Mishler agrees that "research would have gone on, but it would not have made the progress it did." Preuss taps federal agencies for greasing the skids. "Some of these things are initiated by granting incentives, so I think it's wise. It's good to stir the pot and mix people together." Adds Machi Dilworth of NSF, "we would like to foster communication among scientists, to advance science through collaboration and coordination."

References:

1. B. A. Palevitz, "Discovering relatives in the flowering plant family tree," *The Scientist*, 13[24]:12, Dec. 6, 1999.

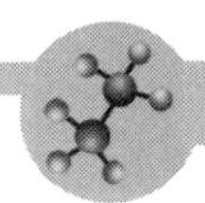

2. L. M. Bowe et al., "Phylogeny of seed plants based on all three genomic compartments: extant gymnosperms are monophyletic and Gnetales' closest relatives are conifers," Proceedings of the National Academy of Sciences, 97:4092-97, April 11, 2000.
3. K. M. Pryer et al., "Horsetails and ferns are a monophyletic group and the closest living relatives to seed plants," *Nature*, 409:618-22, Feb. 2, 2001.
4. B.A. Palevitz, "Arabidopsis genome. Completed project opens new doors for plant biologists," *The Scientist*, 15[1]: 1, Jan. 8, 2001.
5. B. A. Palevitz, "Rice genome gets a boost," *The Scientist*, 14[9]:1, May 1, 2000.

4 Discovering the Truth About Dinosaurs

Discoveries made in the last few years are causing scientists to rethink the classification of various types of dinosaurs. One of the most basic questions is whether dinosaurs were warm-blooded or cold-blooded. Despite studying dinosaurs for decades, taxonomists have not been able to answer definitively this question. The following article, "Were Dinosaurs Cold Blooded?" explores the debate among scientists who are trying to establish the characteristics of dinosaurs. Some scientists believe that dinosaurs as a species flourished so successfully because they were warm-blooded. Meanwhile, others believe that physiological evidence supports the idea that they were cold-blooded.

It is particularly difficult to establish the characteristics of an extinct species since it cannot be studied directly. However, understanding such species is important to understanding the evolutionary relationships among animals living today. "Were Dinosaurs Cold Blooded?" looks at some

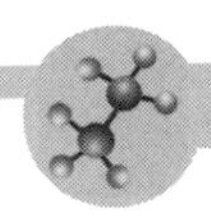

of the ways that scientists are attempting to answer such questions, for example, by comparing features of modern animals with those of dinosaur fossils, and by using modern techniques such as computed tomography (CT) scans. —JF

"Were Dinosaurs Cold Blooded?"
by Brendan I. Koerner
***U.S. News & World Report*, August 18, 1997**

Abstract: Paleontologists hold differing opinions about how dinosaurs lived, and the division is represented in film representations of the animals. Some believe that they were warm-blooded, which explains their evolutionary success, while others believe physiological evidence exists that they were cold-blooded.

The dinosaurs most of us over the age of 20 grew up with were plodding beasts with pea-size brains. In textbooks and schlocky B films, they were portrayed as little more than souped-up crocodiles, lurching lethargically about on splayed-out legs, hunched over like Quasimodo. Like the modern-day reptiles they were thought to resemble, dinosaurs were cold blooded: unable to self-regulate their body temperatures and dependent on the sun alone for warmth.

The budding paleontologists of today's kindergarten set are being raised on a very different crop of "terrible lizards." Bipedal carnivores, clever and fleet-footed, zip around children's literature in voracious packs. Ninety-foot-long sauropods gracefully rear up on their hind legs

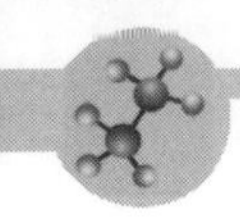

in coloring books. And the fierce velociraptors of Jurassic Park are able to fog up a window with their steamy breath—a sure-fire sign of a warm-blooded animal's ability to regulate its internal thermostat under almost any condition.

It is that last revisionist detail that has divided the paleontological world into rival camps. For some, endothermy, the scientific name for warm bloodedness, is the only way to explain the dinosaurs' evolutionary success. Without the ability to keep their bodies at optimum temperatures regardless of their surroundings, they argue, dinosaurs could never have dominated the globe for 160 million years.

Skeptics counter that ectothermy, the proper label for cold bloodedness, was the logical strategy for dinosaurs living in the Mesozoic Era's generally sweltering heat—and, this group claims, the only option that is supported by physiological, rather than circumstantial, evidence.

The revisionist view that has so captured the public imagination has long been led by Robert Bakker, a former evangelical preacher who has defended dinosaur warm bloodedness with sermonlike intensity. As a Yale undergraduate in the late 1960s, he assisted the legendary paleontologist John Ostrom in his landmark research on Deinonychus, an agile carnivore whose sleek skeleton seemed built for a life of speed more befitting a warm-blooded bird than a cold-blooded reptile. Bakker went on to become paleontology's enfant terrible, a crusader against slow-moving, dimwitted, crocodilian dinosaurs. He proposed such self-described "heretical" ideas as a 10-ton triceratops that could gallop past a charging rhino, and brontosaurs that gave birth to live, 500-kg [1,102 lb] young.

Above all, he painted a picture of dinosaurs that were every bit as endothermic as humans, who manage to keep their body temperature around 98.6 degrees Fahrenheit night and day, winter and summer. Instead of spending their days lazily basking in the sun and occasionally trudging along at a torpid pace, Bakker's dinosaurs—which he wryly termed "nature's special effects"—moved at constant speeds, their postures fully erect in the manner of birds and mammals. "Meat-eating dinosaurs related to *Tyrannosaurus rex* cruised at 3 to 4 miles an hour," claims Bakker, who bases his conclusion on fossilized footprints. "No turtle anywhere cruises at 3 to 4 miles an hour."

Bakker and his acolytes also point to dinosaurs' relatively fast growth as evidence of endothermy. Mammals and birds, which develop quickly compared with ectothermic reptiles, have bones characterized by microscopic channels that appear complex and crystal-like under the microscope. These elegant patterns form when growing bone meets and meshes with connective tissue, capturing blood vessels in dense, woven structures called Haversian canals. Armand de Ricqles, a University of Paris anatomist, found that dinosaur bones exhibited those same intricate channels rather than the simpler, less dense structures common to reptiles. "We see the same well-vascularized bone in mammals but not in turtles and crocodiles," says Kevin Padian, a paleontologist at the University of California, Berkeley. "The way the

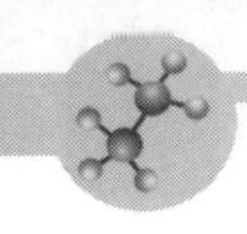

bones grew, dinosaurs seem to have been active all the time." That pace of activity, argue Bakker and his cohorts, is the telltale sign of warm bloodedness.

With Bakker's charisma and de Ricqles's bone histology work, endothermic dinosaurs quickly became the rage. Books were revised, natural-history museums scrambled to accommodate the shift, and Bakker became a dinosaur superstar, commanding speaking fees of up to $10,000.

Feed me. Although the public fell head over heels for the warm-blooded dinosaurs, many within the scientific community remain wary of Bakker's claims. Since measurements show that endotherms require up to 20 times more food than ectotherms, some question how the gigantic dinosaurs could possibly have eaten enough if they were warm blooded. "Can you imagine if a herd of brontosaurs were endothermic?" asks Frank Paladino, a physiologist at Indiana-Purdue University. "They would have eaten through North America in a couple of weeks." The problem would have been worse for endothermic carnivores, for, as James Farlow of Indiana-Purdue notes, "there's a lot less meat on the hoof than plant on the stem."

Bakker has tried to explain away this apparent shortcoming by asserting that predators were very rare and thus able to feast on ample prey. But, as Paul Sereno of the University of Chicago notes, an incomplete fossil record has made it "very, very difficult to reconstruct the number of predators and prey."

The evidence based on bone structures has come under fire, too. Tomasz Owerkowicz, a young Harvard

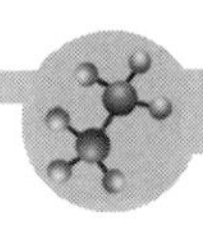

University researcher, has asserted that the dense canals that de Ricqles detected could have resulted from physical exertion rather than endothermy. In an ingenious experiment, Owerkowicz gave cold-blooded monitor lizards regular treadmill workouts and then compared their bones with those of nonaerobicized contemporaries. The well-exercised group showed the same kind of complex channels characteristic of mammals, birds, and de Ricqles's dinosaurs, suggesting that Haversian canals are causally linked to an active lifestyle rather than warm bloodedness. South African histologist Anusuya Chinsamy has also countered some of the bone structure argument, contending that dinosaur bones exhibit bands called lines of arrested growth. These are characteristic of modern-day ectotherms, whose growth rate speeds up and slows down according to seasonal temperature fluctuations. Chinsamy concluded that dinosaurs grew at a more reptilian pace than envisioned by the Bakkerites.

Rather than just playing spoilsport, the ectothermic side has sought to boost its case with hard physiological evidence. John Ruben, a physiologist at Oregon State University, believes he may have found the answer in turbinates, tiny whisps of bone or cartilage deep inside the nasal cavities of mammals and birds. These structures make warm bloodedness possible by limiting water loss. When warm, moist air is exhaled, the water condenses on the turbinates; the next breath brings water vapor back into the lungs. "If [endotherms] didn't have respiratory turbinates, there is no way they could lose that much water" and survive, says Terry Jones, one of Ruben's assistants. Turbinates have never been found in living ectotherms—nor in dinosaurs.

Bet on the croc. Although Ruben's team believes they finally have the proof to cool down dinosaurs for good, they deny that they're trying to drag the animals back into lethargy. "Cold blooded doesn't necessarily mean slow and sluggish," says Jones. The Komodo dragon, the world's largest living lizard, hunts deer. "And deer are pretty active," he says. Paladino agrees: "Ectotherms can do some pretty amazing things," he says. "If I put you on a beach with a 15-foot crocodile and you try to get away, I'll put my 10 bucks on the crocodile."

Many on the cold-blooded side now use the term "gigantothermy" to describe the unique energetics of large dinosaurs. Being huge is one way to maintain a relatively constant body temperature despite cold bloodedness: Large things—which have a lot of bulk in relation to their skin area—lose heat to the outside world much more slowly than do small things. Had they been endothermic, argues James Spotila, a biologist at Drexel University, the large dinosaurs would have experienced a "meltdown," as they would be unable to dissipate internally generated heat at a fast enough rate. However, if they were indeed cold blooded, the slow heat loss associated with gigantothermy would allow them to stay relatively warm—and thus avoid a reptilian torpor—when confronted by the night or an overcast day.

In the generally tropical climate of the Mesozoic, ectothermy may have given dinosaurs an edge over

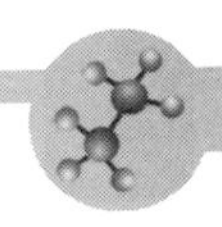

warm-blooded mammals, which had to spend a great deal of energy thermoregulating themselves. Since ectotherms require so much less energy than do birds and mammals, "it's a very, very nice way to make a living if you're in an equitable climate," says Ruben. Contrary to the popular belief that warm bloodedness is always the superior strategy, ectothermy might have been key to the dinosaurs' long reign. Saying that endothermy is superior, says Peter Dodson, a paleontologist at the University of Pennsylvania, is just evolutionary "chauvinism."

The warm-blooded camp, however, is unconvinced by the new set of evidence. Bakker says that Ruben's turbinate research doesn't take into account the possibility that dinosaurs could have utilized an alternative, as-yet-unknown structure to limit water loss. "Ruben's argument is like an expert on piston-driven airplanes looking at a jet and saying you don't have a propeller," he says. Berkeley's Padian, who notes that "behavior precedes hardware in evolution," says dinosaurs may have managed warm bloodedness using mechanisms far different from those found in contemporary animals. Bakker believes that chambers found in Tyrannosaurus skulls may have acted as water-loss regulators in place of nasal turbinates.

Bakker also points to fossils that have been found in Alaska and Australia—two of the very few Mesozoic locales where the mercury occasionally dipped below freezing—as chinks in the seemingly ironclad case for ectothermy. "You don't have Komodo dragons in Seattle, walking into Starbucks," he says. Adverse weather would have particularly affected the smallest of dinosaurs—some of which ranged down to chicken size—who couldn't limit their heat loss through gigantothermy. Cold-blooded advocates have contended that hibernation or migration would have been viable alternatives, but those explanations remain in the realm of conjecture.

Unless time machines or Jurassic Park's DNA cloning technique miraculously become realities, the controversy can never be definitively resolved. "I would never say we know for sure, because we can't," admits Ruben. But although the debate will probably never end, there is little doubt as to which side has more ominous implications for our own species: If dinosaurs were indeed endothermic, then their sudden disappearance 65 million years ago may bode ill for a human race that seems to consider itself invincible. "Maybe we have to rethink our nonvulnerability to global change," explains William Showers, a geochemist at North Carolina State University. "We can't take comfort in being warm blooded if the dinosaurs were warm blooded, too."

What's in a Nose

Turbinates—wispy bones found in the nasal cavity of mammals and birds, but not in reptiles and dinosaurs—support the theory that dinosaurs were cold blooded.

Dog. Warm-blooded animals breathe rapidly and would quickly dehydrate if they didn't recycle water in their breath. Turbinates capture moisture as the animal exhales.

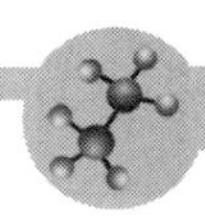

Alligator. Because cold-blooded animals, have a lower metabolic rate, they breathe more slowly and can manage without turbinates.

Dinosaur. CT scans show that dinosaurs lack the telltale ridges that turbinates leave behind in fossil remains.

One of the more popular theories among taxonomists and paleontologists has been the idea that birds are descended from dinosaurs. This article, "Four-Winged Dino Stuns Fossil Hunters: The Discovery of a Dinosaur with 'Extra Wings' Could Force Us to Rethink the Origins of Bird Flight," discusses recent discoveries about dinosaurs in China. These include fossils of a small dinosaur that had flight feathers on its legs, as well as its tail and arms. This would give it a unique configuration with two sets of wings (arms and legs)—something never before seen by paleontologists.

This configuration could mean that not all birds took to the sky in the same way, and in turn, that not all birds are descended from the same dinosaur ancestor. This article not only sheds light on new discoveries regarding dinosaurs themselves, but also illustrates the type of difficulties inherent in trying to construct a hierarchy of related organisms. —JF

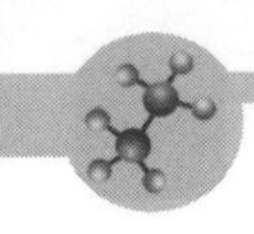

"Four-Winged Dino Stuns Fossil Hunters: The Discovery of a Dinosaur with 'Extra Wings' Could Force Us to Rethink the Origins of Bird Flight."

by Jeff Hecht
***New Scientist*, January 25, 2003**

A stunning set of six fossils discovered in China looks set to overturn our ideas about how birds first took to the sky. The fossils show a small dinosaur that had flight feathers covering its legs, as well as its tail and arms, forming an extra pair of wings never before seen by palaeontologists.

The announcement comes days after a scientist in the US published details showing that young partridges flap their wings to improve their grip on steep slopes, which may explain why wings evolved. Together, the two discoveries may represent a turning point in the contentious study of avian evolution.

"We need to be prepared to change some cherished notions," palaeontologist Larry Witmer of Ohio University told *New Scientist*. Traditionally, experts have been split between two opposing theories: either flight began with small, fleet predatory dinosaurs leaping from the ground into the air, or by different animals jumping to earth from trees. The new studies suggest the picture is more complex, with the ancestors of birds potentially finding numerous ways to take to the sky.

The new specimens "are potentially as important as Archaeopteryx," the famous feathered fossil discovered in the 1860s that alerted scientists to the link between

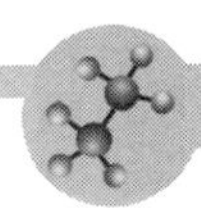

dinosaurs and birds, says Kevin Padian of the University of California at Berkeley. They are of a small dinosaur belonging to the Micro raptor genus, the most primitive of the two-legged predatory dinosaurs called dromaeosaurs, which are known to be closely related to birds. Earlier Microraptor fossils lacked feathers, but the Chinese specimens, which appear to belong to a new species, have the most extensive coat of feathers ever seen on a dinosaur.

The best-preserved skeleton is just 77 centimetres from the nose to the tip of its tail, making it light enough to fly. There is "no doubt the new animal is a flying animal," says Xing Xu from the Institute of Vertebrate Paleontology and Paleoanthropology in Beijing, who details the new fossils in *Nature* (vol 421, p335).

Xu describes the new Microraptor as a "four-winged dinosaur," which he says was not capable of powered flight. Instead it used all four limbs to climb trees, and then glided back down. But *Microraptor gui*, as it has been called, is unusual because the feathers at the ends of its arms and legs are twice as long as those close to its body. In most creatures, gliding surfaces are narrower the further they are from the body.

The feathers on *M. gui*'s legs are particularly baffling. Xu thinks *M. gui* spread its back legs to generate lift, but other dinosaurs could not do this. "This is so far out of the box that we need to sit back and figure out how this can work," says Witmer.

Palaeontologists have had more time to digest the baby partridge findings, which Ken Dial of the

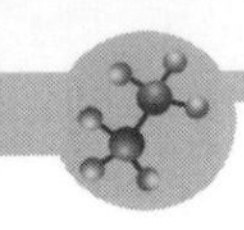

University of Montana had presented at meetings before publishing the details in *Science* (vol 299, p 402). Dial sought to answer a central question about the origin of flight: why did the ancestors of birds evolve wings that would initially have been too small to use for flight?

Dial had a hunch. One-day-old chukar partridges are able to run to avoid predators, and Dial's son had noticed the chicks rapidly flapped their wings as they ran up bales of hay. Could this be helping them climb, Dial wondered?

He clipped or plucked the wing feathers of some birds, then used high-speed photography to see how they fared climbing slopes of different inclines compared with birds with intact wings. He found that even chicks with intact wings ran rather than flew up the slope. While doing so they flapped their wings from head to foot, a stroke that is 90 [degrees] out from the back-to-belly stroke of normal flight.

Dial found that the bigger the chicks' wings, the steeper the slope they could climb. The wing motion was helping to push the birds down, so their claws grasped rough surfaces more firmly. The birds run so fast no one had noticed the flapping before.

The earliest feathered dinosaurs may have behaved in a similar way, Dial says. He suggests that baby dinosaurs tried to escape predators by running to higher ground, seeking refuge up trees for instance. Even the smallest wings would have provided an edge in this race for survival. Further increases in wing size could then have led to gliding and powered flight.

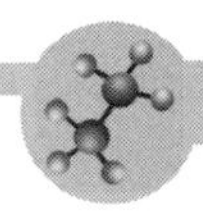

Luis Chiappe of the Los Angeles County Museum of Natural History says Dial's experiments are "impeccable." But he points out we may never know if baby dinosaurs behaved like partridges.

The ancestry of the first birds also remains an open question. M. gui lived around 130 million years ago, at least 20 million years after Archaeopteryx, the first known bird. So was it a living fossil in its time, a throwback to a common ancestor that never developed powered flight, or a bizarre evolutionary experiment, a dead-end twig on the complex family tree of dinosaurs and birds? Perhaps it was something else altogether. Palaeontologists are still scratching their heads.

Recent discoveries in the world of dinosaurs indicate that our entire understanding of the types of dinosaurs that existed in prehistoric times may be inaccurate. This article, "Here Be Monster," describes work by scientists in South America who have discovered that dinosaurs in the southern hemisphere may have been quite different from those in North America.

Across South America, dinosaur hunters are digging up creatures from the Cretaceous period—the third and final part of the dinosaur era, from 65 to 144 million years ago. The problem is that these creatures are very different

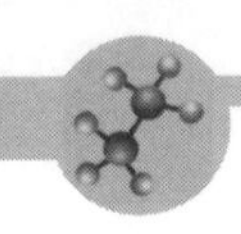

from those "typical" dinosaurs and other animals of the period with which we are familiar.

Since the work on North American dinosaurs forms the basis of our present classification of the types of dinosaurs that once existed, the new information being uncovered in South America may require scientists to reevaluate what types of dinosaurs once walked on Earth and how the different types of dinosaurs were related. This article discusses these revolutionary new discoveries, explains how dinosaurs were traditionally classified and how they might have to be reclassified. As well, it examines what such information could tell us about life in Earth's distant past. —JF

"Here Be Monster"
by Graham Lawton
***New Scientist*, September 23, 2000**

The icons of the dinosaur age are about to be knocked off their pedestals. Graham Lawton goes in search of T. rex's *successors.*

If it's September, then it must be time for Paul Sereno to leave civilisation behind and go looking for dinosaurs. This year, the University of Chicago palaeontologist is contemplating four months in Niger's Tenere desert, a vast arc of wilderness stretching from the Algerian border to the heart of the Sahara. Tenere is so remote that even geographers call it "the desert within a

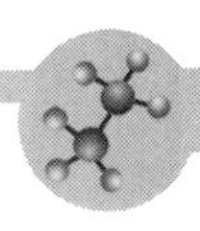

desert." To get there, Sereno's team will have to strike out across open dunes in four-wheel-drive trucks, tracking their progress by global positioning satellites and hauling their water supply behind them. It's hardly well-worn dinosaur territory, but that's the point. Sereno isn't looking for well-worn dinosaurs. He's hoping for oddballs, like the three he dug up last time he went to Niger. And he reckons he'll find them, no trouble.

Sereno is not alone. From South America to Madagascar, dinosaur hunters are unearthing creatures from the Cretaceous period—the third and final part of the dinosaur era, from 144 to 65 million years ago—that are startlingly different from those we have come to think of as "normal." The animals they're digging up—and the ones that don't seem to be there to be dug up—are giving us a much clearer idea of what the world was like as the age of the dinosaurs drew to a close. And it turns out that what most of us think of as classic Cretaceous dinosaurs were actually nothing of the sort. In fact, they were regional specialities, confined to a small corner of the northern hemisphere. It's as if today's zoologists had focused on Australia and concluded that kangaroos and koalas were the dominant forms of non-human life on Earth.

Open a child's dinosaur book and chances are you'll be presented with three world views, corresponding to the three great divisions of the dinosaur age. First comes the Triassic, with primitive monsters such as Plateosaurus and sail-backed, snaggle-toothed Dimetrodon (not actually a dinosaur, but a mammal-like reptile). Then there's the Jurassic and its familiar beasts: giant, long-necked

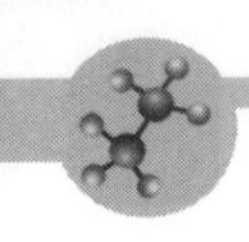

sauropods like Diplodocus, Apatosaurus (formerly Brontosaurus) and Brachiosaurus, plate-backed plant eaters like Stegosaurus and the 12-metre predator Allosaurus. Last of all the Cretaceous, the climax of the era and home to the icons of the age: *Tyrannosaurus* rex and Triceratops, sickle-clawed Velociraptor, snorkel-crested Parasaurolophus, club-tailed Ankylosaurus and duck-billed Hadrosaurus.

But these old friends may soon have to step aside. According to researchers like Sereno, the standard view of the Cretaceous is too regionalist, being based entirely on dinosaurs discovered in North America, Mongolia and a few other parts of the northern hemisphere. During the past ten years, he and a group of like-minded researchers have—literally—broken new ground all over the southern continents in search of the hemisphere's lost dinosaurs. Thanks to their efforts, palaeontologists can begin making generalisations about life in these less explored regions. And some of their conclusions are startling.

For one thing, the southern hemisphere seems to have been a land of giants. Africa, for example, was home to a predator larger than *T. rex*. Argentina had two or three others that were even bigger. Their prey was truly gargantuan: one sauropod from Argentina was the largest land animal ever to walk the Earth.

In other respects, the south was an evolutionary backwater. While life in the northern hemisphere exploded into myriad forms, the southern continents plodded on as normal. "Something extraordinary happened in the north at the beginning of the

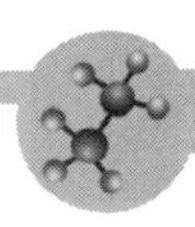

Cretaceous," says Philip Currie, director of the Royal Tyrrell Museum of Palaeontology in Drumheller, Canada, who has spent time hunting for dinosaurs in Patagonia. "Things happened much more gradually in the southern hemisphere. The dinosaurs there got more sophisticated, but basically we have an extension of the Jurassic."

"It's a divided world," adds Sereno. "There are clear differences between north and south." He should know. The dinosaurs he discovered in Niger in 1997 were so unusual they took more than two years to sort out. First there was Suchomimus, a crocodile-jawed, sail-backed monster that fished the rivers of West Africa 100 million years ago. Then there was Jobaria, a primitive, long-necked giant that was already a "living fossil" when it roamed the swamps 135 million years ago, since it looked 40 million years older. And last of all Nigersaurus, an enigmatic spade-headed herbivore with hundreds of teeth that's so odd-looking even Sereno describes it as "marvellously bizarre." What's more, the region seems devoid of the archetypal Cretaceous animals. It's as if Sereno had discovered a lost world in the middle of the Sahara.

And in a sense it is a lost world, at least when viewed from the northern hemisphere. For most of the dinosaur era, the world's landmasses were cemented together in a massive supercontinent called Pangaea. Animals that evolved in one part of the world could—and did—spread quickly to others. Bones of the Jurassic predator Allosaurus, for example, have been found all over the world, from the US to Portugal, Australia and Tanzania.

Jurassic Shift

But during the late Jurassic, the world started to change. Continental drift split Pangaea in two and by the early Cretaceous, 140 million years ago, the world's dry land was divided into two continents. To the north lay Laurasia (which is now North America and Eurasia) and to the south, Gondwana (South America, Africa, Madagascar, India, Australia and Antarctica). Rifts continued to develop throughout the Cretaceous, dividing Laurasia in two and splitting the Gondwanan landmass into smaller and smaller chunks.

This fragmentation, of course, stopped dinosaurs from wandering freely all over the world. And that ought to mean that different types of animals evolved in different regions. The most pronounced division should be between Laurasia and Gondwana. Palaeontologists have long acknowledged that this should be the case, but had little evidence to back it up. "There was a feeling that the southern hemisphere was different," says Dale Russell, the senior curator of palaeontology at the North Carolina State Museum of Natural Sciences in Raleigh, and another Niger veteran. "But the record was very scrappy." Most fossil hunters were content to focus on the rich Cretaceous deposits in Asiamerica, today's western North America and East Asia. Few bothered with the southern hemisphere. As recently as 10 years ago, the five best-known dinosaur faunas all came from Laurasian countries (the US, Mongolia, China, Canada and Britain).

The revision of this world view began in earnest in 1985 when Argentinian palaeontologist Jose Bonaparte

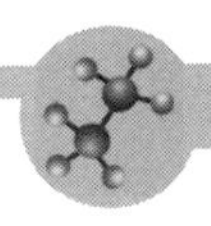

unearthed two new flesh-eating dinosaurs in Patagonia. One, Carnotaurus, was a superb specimen, an entire skeleton so well preserved that its skin had left impressions in the surrounding rock. It came from the middle Cretaceous, around 100 million years ago. Yet it was unlike anything ever discovered from that time. For one thing, it had horns above its eyes—hence the name, which means "meat-eating bull." The shape of its brain case—one of the prime diagnostic features of predatory dinosaurs—was unfamiliar. Overall, it looked remarkably primitive. The other species, called Abelisaurus, was incomplete, but it had clear similarities to Carnotaurus. Taken together, the discoveries seemed to suggest that there was an unknown group of bipedal predators roaming Argentina around the time *T. rex* dominated the northern hemisphere. In fact, Carnotaurus and Abelisaurus were so unusual that palaeontologists assigned them to a wholly new group of dinosaurs, the abelisaurids, part of a lineage that diverged from the main branch of carnivorous dinosaurs around 230 million years ago. More abelisaurids later turned up in India and Madagascar.

Southern Titans

Patagonia, meanwhile, continued to yield enigmatic monsters. In 1991, Bonaparte dug up a titanosaur, a long-necked sauropod herbivore around 24 metres in length. Two years later, he found another species. Though fragmentary, its remains indicated that it was 45 metres long and weighed 100 tonnes, making it the largest animal ever to have walked the Earth. He and

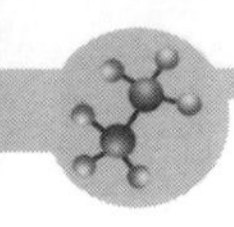

his colleague, Rodolfo Coria of the Carmen Funes Museum in Neuquen, called it Argentinosaurus.

Coria himself was also finding spectacular new species. In 1995, this time working with Leonardo Salgado of the Museum of Natural Sciences in Neuquen, he discovered a meat eater that brought Patagonia's dinosaurs to the world's attention. It looked a lot like *T. rex*, but it was bigger. Giganotosaurus, as they called it, was around 14 metres long and weighed 8 tonnes, making it bigger than even the biggest *T. rex*, which up to that point had been thought of as the largest predator the world had ever seen. But Giganotosaurus was no match for *T. rex* in one respect: its brain was about half the size. In March of this year, Coria and Philip Currie announced the discovery of another meat eater that was bigger still. Currie says there are remains of a third Giganotosaurus-like species that dwarfs the lot.

The discoveries add up to an inescapable conclusion: in South America, Cretaceous dinosaurs tended to be bigger, dumber, and more primitive than their northern contemporaries. The most abundant herbivores were long-necked sauropods, a group that died out over most of Laurasia around the beginning of the Cretaceous. The most common predators, meanwhile, were the abelisaurids, dinosaurs that seem to have flourished in isolation on the southern continent and are absent further north. And right at the top of the food chain were monstrous, pea-brained beasts like Giganotosaurus—throwbacks to the allosaurs of an earlier age. "Most South American dinosaur fauna are

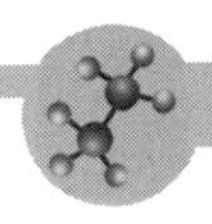

oversized forms," says Coria. "They represent primitive dinosaurs that were widely distributed around the world during the Jurassic period, but survived another 50 million years into the Cretaceous in South America. This was their last bastion before they went extinct."

Many of the patterns found in Patagonia are repeated elsewhere. Abelisaurids, for example, have turned up in India, Madagascar and Africa. And titanosaurs have now been found all across the southern hemisphere. "Everywhere other than Asiamerica, sauropods are easily the most common dinosaurs," says Thomas Holtz, a palaeontologist at the University of Maryland. That's especially true in Africa. Apart from the titanosaurs, the continent has at least two other groups of sauropod, both discovered in Niger by Sereno. One is Jobaria, the herbivorous "living fossil." The other is Nigersaurus, the spade-headed weirdo. Though Nigersaurus is descended from Diplodocus, the gangly Jurassic sauropod whose skeleton graces the foyer of London's Natural History Museum, it looks very different. It's one of the smallest sauropods on record, reaching only 15 metres in length. And its mouth is stuffed with rows and rows of teeth—as many as 600 per individual. Nigersaurus may have filled the same ecological niche as the duck-billed dinosaurs of Asiamerica. And, according to Sereno, there are hints of a similar species in South America.

Nigersaurus isn't the only odd African herbivore. In 1999, Russell published details of an eccentric creature from Niger called Lurdusaurus ("weighty lizard"). "It's a funny thing," he says. "It's like a hippopotamus, low

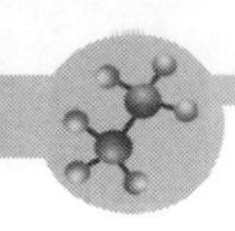

to the ground, barrel-chested and with a small head. And it's got this powerful thumb claw." Not, in other words, a lot like your regular northern plant-eater.

Africa also has big predators that fit the Gondwanan pattern. On their maiden expedition to Niger, in 1993, Sereno's team unearthed the complete skeleton of a meat eater, the best specimen ever found in Africa. They called it Afrovenator ("African hunter") and assigned it to a group called the torvosauroids. These evolved during the Jurassic but, until 1993, were unknown in the Cretaceous. Two years later the team unearthed two more predators, this time in the Kem Kem region of Morocco. One, Deltadromeus ("delta runner"), was a swift and graceful hunter that seems to have evolved in isolation in Africa. The other, Carcharodontosaurus ("shark-toothed lizard"), was a flesh-guzzling monster that is related to the giants of South America. Carcharodontosaurus could easily have gone toe-to-toe with Giganotosaurus: it was more than 14 metres long and weighed a gargantuan 8 tonnes.

Crocodile Jaws

More finds continued to surface. In 1997, Sereno found another torvosauroid in Niger, the crocodile-jawed Suchomimus. This was the most common predator of its day. It was at least 11 metres long and had an elongated snout bristling with hook-shaped teeth for snaring fish. In 1998, Russell found traces of a similar creature in Morocco. South America, too, has a crocodile-jawed fish eater, the delightfully named Irritator challengeri. This got its name because the only known skull had been damaged by amateur palaeontologists.

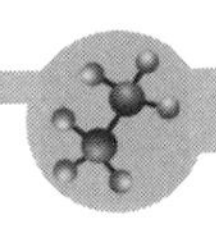

The pictures that are emerging of Africa and South America, then, are strikingly similar. Both were home to giant sauropods and Allosaurus-like predators long after these groups had started to dwindle in Laurasia. And both had eccentric-looking species that evolved in isolation following the split of Pangaea. "You have two sorts of animal," says Sereno. "There are animals that survived only on the southern continents, and there are animals that evolved on the southern continent and are not found elsewhere."

Laurasian Echoes

Within this broad pattern, however, there are some stray threads, The dinosaurs of Australia and Antarctica seem much more like those found in Laurasia. In 1998, for example, a tooth from a duck-billed dinosaur was found in Cretaceous deposits in the Antarctic. "It's a bit of an enigma," says Currie. "Some of the material down there is not like Gondwanan animals. There are some things that are more suggestive of the northern hemisphere." Gondwanan animals have also started showing up in Laurasia. Suchomimus, for example, has a close European relative called Baryonyx, found in early Cretaceous rocks on the Isle of Wight, off England's south coast. The two are so similar that that they may in fact be the same species. What's more, titanosaurs have been uncovered in Western Europe, North America and Mongolia.

"It's a complex picture," says Sereno. "The break-up of Pangaea didn't create insurmountable obstacles to these intrepid explorers." One possible explanation is

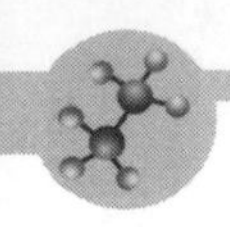

that land bridges formed from time to time as a result of changes in sea level, temporarily reuniting divided land masses and allowing migration and intermingling of species. Another possibility is that the fossil record simply isn't good enough to give us a complete picture.

Whatever the real reason, Sereno and his fellow explorers are sure that the fossil beds of Gondwana will continue to yield surprises. "There's lots more to come," says Russell. They're also convinced that what they find will continue to marginalise *T. rex*, Triceratops and the other Laurasian icons. "The dinosaurs of North America and Eurasia were an unusual endemic fauna, real weirdos that were generated in isolation," says Holtz. "Southern hemisphere dinosaurs are the main strand of dinosaur history."

5 Who Were Our Ancestors?

DNA analysis is providing valuable information that can be used to answer the question of whether or not the two major groups of Stone Age Homo sapiens*—Neanderthals and Cro-Magnons—were related. The following article, "Ancient DNA Enters Humanity's Heritage," provides valuable insight into whether the Cro-Magnons, who appear paleontologically to have replaced the Neanderthals, were descended from them or were a separate unrelated branch of the evolutionary tree.*

Using mitochondrial DNA (which comes from the mother) from two anatomically Cro-Magnon specimens found in a cave in southern Italy, Giorgio Bertorelle of the University of Ferrara in Italy (and his colleagues) have established that Cro-Magnon DNA contains chemical sequences that are similar to those of people today, but which are very different from those found in

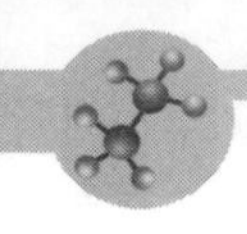

Neanderthal specimens. Such findings have implications for how different groups of human beings developed in different places. This article looks at Bertorelle's findings, the implications of the use of such technology to establish evolutionary relationships among groups of people, and some of the questions that this research has raised. —JF

"Ancient DNA Enters Humanity's Heritage (Stone Age Genetics)"
by B. Bower
***Science News*, May 17, 2003**

Genetic material that Italian researchers extracted from the bones of European Stone Age Homo sapiens, sometimes called Cro-Magnons, bolsters the theory that people evolved independently of Neandertals, the team proposes.

Fossils of two anatomically modern H. sapiens found in a southern Italian cave yielded mitochondrial DNA, which is inherited from the mother, say Giorgio Bertorelle of the University of Ferrara in Italy and his colleagues. The DNA contains chemical sequences that resemble those of people today but differ substantially from those previously isolated from four Neandertal specimens, the scientists report.

One of the Italian Cro-Magnons dates to 25,000 years ago; the other, to 23,000 years ago. Neandertal fossils that have yielded mitochondrial DNA range from about 29,000 to 42,000 years old (SN: 4/1/00, p. 213).

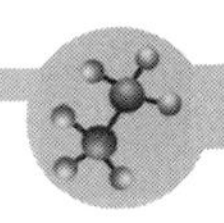

"These results are at odds with the view [that] Neandertals were genetically related with the anatomically modern ancestors of current Europeans or contributed to the present-day human gene pool," Bertorelle's group concludes.

Contamination of ancient DNA can occur easily. However, the mitochondrial DNA obtained from the Cro-Magnon bones exhibits no trace of genetic material from other animals unearthed in the Italian cave or from people who have handled the bones, the scientists assert in an upcoming Proceedings of the National Academy of Sciences.

The researchers compared Cro-Magnon genetic sequences from an especially variable stretch of mitochondrial DNA with corresponding sequences from Neandertal fossils and from 80 people now living in Europe or western Asia.

Cro-Magnon sequences fall within a genetic category shared by people today but not by Neandertals, the scientists report. This result aligns with the theory that modern H. sapiens originated in Africa around 150,000 years ago and then replaced Neandertals in Europe rather than interbred with them, Bertorelle and his coworkers say.

Mark Stoneking of the Max Planck Institute for Evolutionary Anthropology in Leipzig, Germany, an advocate of this single-origin model of human evolution, nonetheless regards the new evidence with caution. He hasn't seen the report but worries that the Cro-Magnon DNA is contaminated. However, mitochondrial DNA analyses of living people align

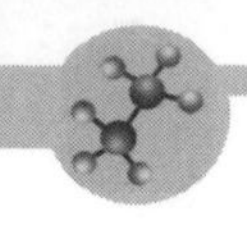

with the single-origin, or out-of-Africa, scenario, Stoneking says.

Adherents of the contrasting multiregional-origin theory of evolution view the Cro-Magnon findings even more skeptically. They argue that anatomically variable H. sapiens in Europe, Africa, and Asia interbred enough over the nuclear DNA—the DNA that holds most of a person's genes indicate that interbreeding of H. sapiens and other Stone Age Asian or European groups, if not Neandertals, contributed to modern humanity's evolution, remarks Alan R. Templeton of Washington University in St. Louis.

DNA analysis is not the only modern biological technique that can be applied to classifying our prehistoric ancestors. Scientists are using 3-D computer graphics technology to attempt to answer questions about ancient hominids. For example, recently there has been a great deal of debate about the nature of Neandertals. Did they become extinct? Were they assimilated?

This article discusses a number of different types of imaging technologies and the ways that scientists are using them to help establish the relationships between prehistoric human beings. Scientists who now use such technology

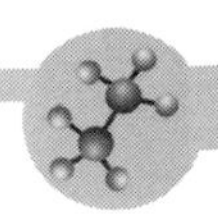

on recent finds also use it to reexamine older fossil finds. This gives them the ability to confirm or reconsider assumptions made in the past.

Using medical imaging technology—such as computed tomography (CT) scans and modern 3-D computer modeling technology—researchers are attempting to establish the anatomical characteristics of prehistoric hominids in a way that will allow us to see how they differed, and thereby, how to classify them. —JF

"The Way We Were: Scientists Make No Bones About the Value of 3D Imaging in Studying Our Distant Ancestors."
by Andy Pasternack
***Computer Graphics World*, September 2002**

They [say] a good man is hard to find. Change that to "a good fossil hominid," and it's just about impossible to find one of those. It's not that there aren't plenty of human fossils in museums. But museum curators would have to have rocks in their heads to part with a rare and fragile fossil skull, even for scientific study. Thanks to computer graphics, however, researchers are learning reams about human origins without having to damage or alter valuable specimens.

A paleoanthropologist's work often requires figuring out how to glean maximum information about human origins from a minimum of fossil fragments. Among the questions scientists ponder are how our

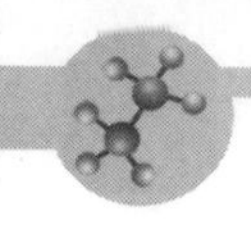

ancestors might have looked, how they got around (by climbing or walking), how they developed and matured, and what their environment was like. Increasingly, these researchers are using techniques such as computer tomography to acquire fossil data; three-dimensional computer graphics to analyze and visualize surfaces, volumes, and internal structures; and rapid prototyping technology to create specimens for further study and public exhibition.

Biologist Christoph Zollikofer and anthropologist Marcia Ponce de Leon, both at the University of Zurich, are among those using computer-aided paleo-anthropology (CAP) to expand our understanding of human origins. In fact, one project brought them face to face with a 30,000- to 50,000-year-old Neanderthal. "Just 10 years ago, research into Neanderthals was considered marginal," says Zollikofer. "In the last five to seven years, there has arisen a lot of debate about them. Did they become extinct? Were they assimilated? It's a very big debate in Europe."

Zollikofer and Ponce de Leon were asked to apply their computer-based techniques and take a new look at a rare set of skull fragments from a Neanderthal child. The Devil's Tower Child, as the fragments have been known, were discovered in 1926 by archaeologist Dorothy Guard in Gibraltar.

Using a CT scanner, the scientists made a full series of one-millimeter serial cross sections of all the fragments. Each pixel in a CT image conveys the object

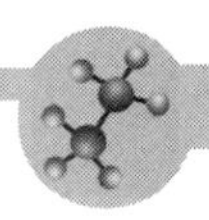

density at that location, allowing thresholding, or the defining of specific object regions. From the CT data, they were able to construct contours to create 3D surfaces. Then, running their own software, called Forces Reconstruction and Morphometry Interactive Toolkit (or FORMIT), on a Silicon Graphics Onyx II, they were able to fit the fragments together into a single skull. "The most exciting moment was when we put those parts together, and they just fit perfectly," Zollikofer says.

Next, the researchers created a stereolithographic replica using a 3D Systems SLA machine. They then morphed data collected from CT scans of modern human children to the Neanderthal skull and sent both the skull replica and the facial imagery to sculptor Elisabeth Daynes in France. "We didn't want to constrain her too much—we wanted to let her follow her own human intuition about how the child should look," Zollikofer says.

Daynes selected eye color, hair color, hair length, and other characteristics, and using the replica and imagery created a statue of a four-year-old that ended up having the facial maturity you'd expect in a six-year-old. This supports the theory that Neanderthals, and other pre-historic humans, developed more quickly than modern humans do, possibly as an adaptation to harsher ice age environments. Conversely, our slower development may have facilitated the development of culture and civilization. "That's completely hypothetical," says Zollikofer. "We don't really

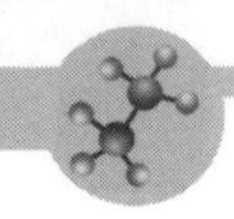

have a clue as to why we develop more slowly." The resulting statue is on exhibit in the National History Museum, London, England.

Out on a Limb

Did our earliest ancestors climb upright or hang suspended from branches? That's one of the questions asked about Proconsul Heseloni, a hominoid that roamed East Africa some 18 million years ago. Many scientists believe Proconsul represents the beginning of the line that ends with human beings. It lived at the exact evolutionary moment when monkeys and apes split into different families, and while Proconsul shares some features with modern apes, such as skull shape and the lack of a tail, it moved more like a monkey. That makes the creature interesting to study, says Dana L. Duren, an anthropologist at Wright State University School of Medicine in Dayton, Ohio.

Duren has been studying the long arm and leg bones of juvenile Proconsuls, using fossil remains found at the National Museums of Kenya. She has focused on the metaphyses, or bone ends. Forces or stresses on the growth plates are believed to affect the pattern of growth, and therefore the morphology of the hones adjacent to the growth plates. "The stresses across the growth plate tell you how it moved," Duren explains. In a human leg bone, the majority of pressure on the knee goes straight down. But a monkey in a tree never extends its knee all the way when it walks, so its bones exhibit different morphology and different stresses, she says.

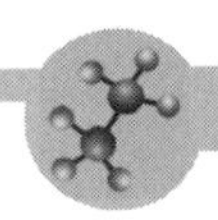

Using 3D graphics, she has been able to resolve the question of how Proconsul got around. First, she made molds of the bone ends with the kind of silicone putty dentists use to take molds of their patients' teeth. From these she made epoxy casts, which were then scanned using a Laser Design 3D scanner. Data from the scans were imported as XYZ coordinates into geographic information systems software, ESRI's ArcView, and its 3D Analyst extension. "ArcView was originally developed to reconstruct and analyze geographic landscapes. But it is ideal for the kinds of analyses I do because to the computer the surface of the metaphyses resembles mountain ranges," says Duren. "The units of measure are just a whole lot smaller."

Once a 3D rendering was created in ArcView, Duren calculated the surface and planimetric areas and other features of the bone ends. "It is virtually impossible to measure surface area without a 3D rendering of these surfaces, especially since some at them are so small," she says. By using both surface area and planimetric area, Duren was then able to compute a roughness index for the bone-end surfaces. When she compared the roughness of the bone end surfaces of Proconsul limbs to that of modern primates, it was clear that the species climbed and moved around on all fours.

Duren is now using ArcView and 3D surfacing to compare humans with other primates and ancient australopithecenes to evaluate the same kind of bone stress patterns. During the last century, only a handful of bone end surfaces were analyzed. "With the GIS

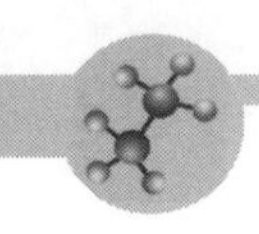

software, you can do a comprehensive analysis," she says. "This technique will open up this area of study."

Mummy Dearest

Remains of long-gone modern humans are also undergoing more in-depth analyses, thanks to 3D graphics. Arthur Andersen—whose Virtual Surfaces computer-aided design shop in Downers Grove, Illinois, specializes in reverse engineering—has brought his skills to bear on behalf of museums that want to electronically reexamine their human specimens. One of these was a 2200-year-old mummy.

Just a few things were known about the mummy. His name was Padu-Hari. Possibly a priest, he died at about the age of 20, around 200 BC. He was found in Akmen, Egypt, from which his remains made their way to the Milwaukee Public Museum in the mid-nineteenth century. When the museum decided to add a high-tech touch to its mummy exhibit, it contracted with Mobil Scanning Laboratories to perform CT scans of the specimens, head to toe. The data for Padu-Hari's head was handed over to Andersen to see what he could do in the way of analyzing and visualizing the skull and face.

Andersen brought the data into Mimics, a PC-based 3D modeling program from Materialise, which allowed him to edit the scan data. The image could be manipulated, and the brightness and contrast could be adjusted as needed. Mimics allowed him to set a threshold for a particular density of pixels, which would then come up as a solid model of the skull. Once the skull was established, he selected less dense material for the mummified skin. "I was working on it at night," he recalls. "I created a threshold and the skull popped right up. I changed the thresholding and up popped his face. I realized I was the first person in 2200 years to look at this guy."

Medical device and supply manufacturer Baxter Corp. loaned its resources to create a prototype of the skull. Baxter's process uses technology from Z Corp., which is different from stereolithography. Z Corp.'s plaster-like material is treated with a curing agent, and there are no internal supports in the prototype, as is the case with stereolithography. Thus, cavities such as the sinuses, which might otherwise have had supports intruding on the structure, are rendered intact.

The mummy skull replica had a few surprises of its own that hadn't shown up in the CT slices. Using Z Corp.'s rapid prototyping technology, Andersen created a physical modal. From there, he was able to establish that the brain had been removed from the skull, a practice common to mummy preparation. The model also revealed the presence of bone chips in the occipital (back) region of the skull. He also discovered that some kind of liquid resin had been poured into the skull, where it dried and crusted.

The experience of using 3D modeling and rapid prototyping to reconstruct mummy remains left Andersen thinking of other ways the technology could benefit paleoanthropologists. For example, researchers

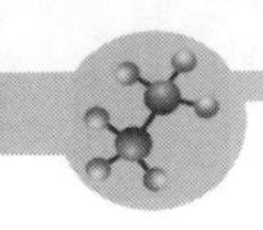

could take 3D models of two skulls, compare the point clouds, and digitally see how they differ, he says. "if you're comparing Paleoindians and Europeans or a Neanderthal and Cro-Magnon, for example, it could really aid in paleontological studies."

That a CAD operator like Andersen is conjuring up paleoanthropological CG applications shows how this technology is transforming the study of human origins. It's also bringing nonscientists into the realm. Zollikofer, who spends most of his time in a multimedia lab, sees great promise in the teaming of CG experts and scientists. He hopes that, eventually, these cross-disciplinary partners will team up to compile 3D quantitative data on the human body. "Normal human variation of anatomy isn't documented in 3D graphics, and it needs to be for these kinds of studies," he says. "That's one of the big projects for the future."

Of course, locating and analyzing fossils has always been a critical part of attempting to classify prehistoric hominids. In this area, as well, new discoveries are continually being made. Recent data on three fossil skulls discovered in Ethiopia revealed that they are the oldest human remains ever found.

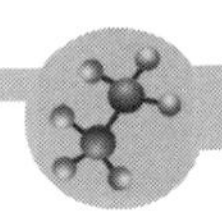

These 160,000-year-old bones date from about the time that our species first appeared. Scientists believe that they may fill a crucial gap in the fossil record. What they have to tell us can help us understand the links among prehistoric peoples. These finds support the theory that early human beings emerged in Africa and subsequently spread to other parts of the world, thus developing into various subgroups. This is a theory with many supporters, but until now, it has only been documented by sketchy physical evidence.

This article provides an excellent overview of the types of techniques that researchers use to establish the age and nature of human fossils. —JF

"The Dawn of Homo Sapiens: Stunning Fossils of the First Modern Humans Are Silencing Debates About Where and When Humanity Made Its Debut."

by James Randerson
***New Scientist*, June 14, 2003**

Three fossil skulls from Ethiopia have turned out to be the oldest human remains ever found. Anthropologists say the 160,000-year-old bones plug a crucial gap in the fossil record around the time that our species first appeared, and give dramatic insights into the lifestyles of the earliest human beings.

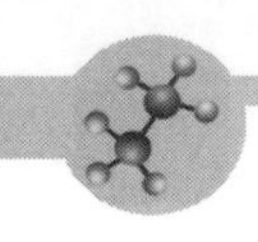

The fossils suggest that the first Homo sapiens ate the bone marrow of hippos, and ritually carried the bones of their ancestors. They are also a dramatic boost to the theory that modern humans evolved in Africa and subsequently spread across the rest of the globe, according to Chris Stringer of London's Natural History Museum, who calls them "landmark finds in unravelling our origins."

News of the fossils' age follows a detailed analysis of the remains, which were found by Tim White of the University of California at Berkeley and his team in 1997 near the village of Herto, 230 kilometres north-east of Addis Ababa (for map, see page 5 of original article). White first stumbled across a fossilised hippo skull protruding from the ground, but the team eventually recovered skull fragments from 10 humans. The site was also littered with stone tools and animal remains.

White's team dated the human fossils with enormous precision using a technique based on the radioactive decay of potassium into argon. When volcanic material is heated it expels argon, but on cooling the gas starts accumulating again in pockets between crystals. By analysing the gas in volcanic debris that settled with the fossils as they formed, the team could fix them between 160,000 and 154,000 years old. This makes them crucial to the debate about where modern humans evolved (see "Where, When, and How," on page 5 of original article).

The prominent "Out of Africa" theory suggests modern humans evolved roughly 100,000 to 200,000

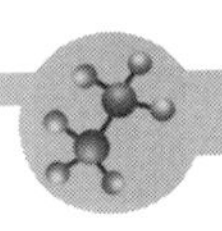

years ago in Africa, then migrated all over the Old World. But there have not been any convincing African human fossils from the time they supposedly appeared. "The problem with the African record is that it has been really sketchy," says White. Now, in a coup for the Out of Africa theory, the Herto fossils place modern-looking humans in the right place at the right time.

White's team painstakingly restored large parts of the skulls of three individuals—two men and a child. More than 200 fragments of the child's skull were strewn over an area of about 400 square metres. It took two years to piece the fragments together, using clues like the impressions made by blood vessels.

The fossils look almost human, but not quite. The skulls have a longer brain case (from front to back) than human skulls, a deeper face and more pronounced brow ridges (*Nature*, vol 423, p 737). So the team have given these people their own subspecies, Homo sapiens idaltu. The subspecies name means "elder" in the language of the Afar people who now live in the Herto region. The skulls are also very large by human standards, suggesting that the adults were hefty individuals. "They'd be a star choice for a rugby player. You'd want them on your team," says White.

The animal fossils near the human remains give intriguing insights into the Herto people's lifestyles. Their African habitat was a warm tropical plain near a shallow freshwater lake, teeming with hippos, crocodiles

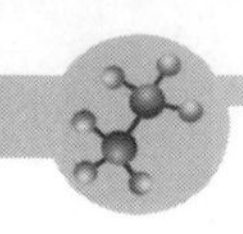

and catfish. The hippo remains suggest the people ate the animals' bone marrow, as the limb bones were deliberately broken. One hippo skull had a gash 10 centimetres long made by a stone tool.

There are also marks on the human skulls. One of the two adult skulls bears parallel grooves carved by a stone tool. This doesn't seem to signal cannibalism because the grooves don't look like the marks made by scraping meat from bones. "There's no meat in the places they're finding the cut marks," adds Sally McBrearty, an expert in stone tools at the University of Connecticut in Storrs.

Marks on the child's skull are even more puzzling. There are cut marks made by very sharp stone flakes in nooks and crannies at the base of the skull. Part of the base of the skull was broken away and the broken edges have been polished. White says this suggests that people carried the skull around—possibly as part of an ancestor-worshipping ritual—and it got smoothed and buffed up in the process. If he is right, this would be the earliest evidence of this kind of cultural trait.

RELATED ARTICLE
"Where, When and How?"

The Herto fossils will have a tremendous impact on a long-standing debate in palaeoanthropology: "When, where and how did modern humans arise?"

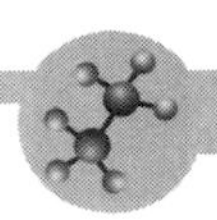

Two competing theories have been slugging it out for more than twenty years. The first is the "multiregional" hypothesis, which argues that llama sapiens arose from llama erectus ancestors in many parts of the Old World simultaneously over a time span of hundreds of thousands of years. Among the most vocal champions of this theory are Milford Wolpoff of the University of Michigan in Ann Arbor and Alan Thorn of the Australian National University in Canberra.

The rival hypothesis, called "Out of Africa," sees things very differently. It says that early modern humans arose just once quite recently, a little more than 100,000 years ago, in a single place, probably Africa. Descendants of that population spread throughout the Old World, replacing more archaic forms. Chris Stringer of London's Natural History Museum is the chief protagonist of this theory.

If the Out of Africa model is correct, then you should expect to find the earliest forms of modern humans in Africa. Candidate fossils up to 300,000 years old have been found in eastern and southern Africa, but they were usually fragmentary and hard to put a date on. These factors gave critics of the hypothesis enough wiggle room to argue that the case was not proven, and that multiregionalism was the preferred interpretation. However, the Herto fossils make it much more difficult to dispute the Out of Africa view. The fossils are beautiful specimens, and they have been accurately dated to 160,000 years old.

H. erectus first appeared in Africa almost two million years ago, and rapidly expanded its range into Asia and, later, Europe. Various forms of so-called archaic sapiens then popped up in the Old World, most famously the Neanderthals in Europe. To the multiregionalists, the Neanderthals were an intermediate form, on the way to truly modern humans in Europe.

But the fact that the Herto people were so strikingly modern long before the "classic" Neanderthals lived in Europe makes that tough to argue. And DNA evidence culled from Neanderthal bones in recent years indicates that they were evolutionary dead ends that contributed little, if anything, to modern humans (*New Scientist*, 17 May, p14).

Moreover, genetic evidence from mitochondrial DNA and other sources has been building inexorably over the past several decades, giving ever stronger support to the Out of Africa model. Most recently, this data has been interpreted to indicate an African origin of about 160,000 years ago. The Herto fossils support that.

Proponents of the multiregional view insist that the new fossils are not the nail in the coffin for their idea. Wolpoff, for instance, argues that although they represent one of the ancestors of modern Europeans, there could well have been others. He told New Scientist that early humans spreading from Africa could have interbred with Neanderthals in Europe and *H. erectus* in Asia to create modern populations.

But proponents of the Out of Africa theory say that the Herto discovery settles the question of "where, when and how?" at the heart of the debate. Although it is always risky in science to say "case closed," that's how it looks.

6 Can We Create a Universal Tree of Life?

All of the efforts of taxonomists and other scientists described thus far in this book are ultimately aimed at one major goal—establishing a complete "tree of life" in which all organisms will be classified and all evolutionary links among them will be clear. Can this goal be achieved? Many scientists think so. By combining the expertise of scientists and technical experts in fields such as DNA analysis, software design, systematics, information systems, and others, they hope to develop the definitive tree of life.

The following article, "Dating the Tree of Life," discusses the problems of establishing exactly where in time various species fit. Should similarities in molecular (genetic) data be used to establish where various organisms fall on the evolutionary tree of life or should morphological (physical similarity) and paleontological data be the guiding factors? The authors of this article argue that (1) more

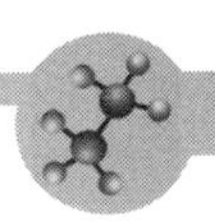

molecular and morphological/paleontological trees and dates agree than disagree, (2) neither paleontological dates nor molecular dates are perfectly accurate, and (3) it is possible to work toward a middle ground where apparently disputed ages eventually converge on an agreed-upon date. —JF

"Dating the Tree of Life"
by Michael J. Benton and Francisco J. Ayala
Science, June 13, 2003

The reconstruction of segments of the "tree of life" has long been a driving force for systematists. Since the mid-1980s, there has been an exponential growth in the number of phylogenetic papers published each year[1]. The tree of life project, whose end point is the construction of the single phylogenetic tree linking all species living and extinct, promises to be a substantial, international research program involving thousands of biologists. The scientific aim is the same as that set out by Darwin:[2] to understand where life came from, the shape of evolution, and the place of humans in nature and to determine the extent of modern biodiversity and where it is threatened.[3, 4]

A key concern in this project is the calibration of phylogenies against time. This surfaced in the 1960s with the first attempts to estimate divergence dates in a phylogenetic tree from molecular evidence. Since then, the value of the molecular and morphological or

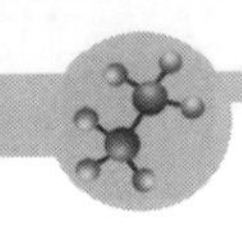

paleontological approaches[5-7] has been recognized. However, some commentators indicate that, in cases of dispute, molecular dates should generally[8-10] or always[11, 12] be preferred. We suggest that there are no such simple solutions. First, more morphological or paleontological and molecular trees and dates agree than disagree. Second, although paleontological dates by definition are always underestimates (providing specimens are correctly identified), it may be that molecular dates are always overestimates. Third, close study and care of calibration points can lead to rapprochement, where apparently disputed ages eventually converge on an agreed date.

Early Origins of Major Clades

In some noted cases, the molecular age estimates for origins of groups are about twice as old as the oldest fossils. The range of molecular estimates for the origin of metazoans is 600 to 1500 million years ago (Ma),[9, 13-16] with many recent estimates narrowing it down to 700 to 1000 Ma.[15, 17-19] There is fossil evidence of Precambrian metazoans but nothing before about 600 Ma. The new molecular consensus, however, is that basal splits among major animal clades happened about 1000 Ma and that the modern phyla, such as molluscs, arthropods, brachiopods, and echinoderms, diverged about 600 to 800 Ma. There are three reasonable explanations for these discrepancies (20): (i) the molecular and paleontological dates may mark different events,[16, 21] for example, the genetic divergence of lineages (molecular date) and

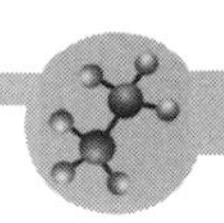

the acquisition of hard skeletons (paleontological date); (ii) the fossil dates could be too young[8, 9, 13] as a result of an absence of fossils from much of the Precambrian, because either they lacked skeletons, they were microscopic, they did not become incorporated into the rocks, or they have been missed by paleontologists; or (iii) the molecular dates could be too old as a result of unaccounted-for variations in the rates of molecular evolution, incorrect calibration points, or inadequate correction for other biases.[14, 22, 23]

The first vascular land plants are found as fossils in the Silurian, and earlier evidence from possible vascular plant spores may extend the range back to the Ordovician, 475 Ma,[24] considerably younger than a molecular estimate of 700 Ma.[25] A similar gap exists for angiosperms, with the oldest generally accepted fossils being from the Early Cretaceous (120 to 130 Ma).[24] DNA sequence evidence places the divergence of angiosperms in the Mid Jurassic, 140 to 190 Ma,[26] but the date could be much older, Carboniferous (290 Ma),[27] if it turns out that the sister group of angiosperms is the gymnosperms.

For modern birds, molecular estimates place the split of basal clades and modern orders at 70 to 120 Ma.[12, 28, 29] Although many supposed Cretaceous representatives of modern bird orders have been cited,[28, 29] most have been disputed, generally because the fossils are isolated elements.[30] The oldest uncontroversial fossils of modern bird orders date from the Paleocene (60 Ma), much younger than most molecular estimates of origins.

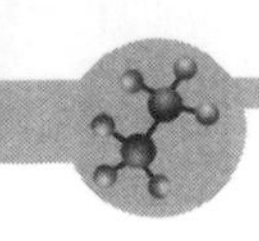

The dating of the radiation of modern placental (eutherian) mammals also seemed to be an example of unusually early molecular dates. The paleontological view[31] is that placentals split from marsupials some time in the Early Cretaceous (144 Ma). The first molecular dates[8, 12, 32] seemed much older: origin of eutherians in the Late Jurassic (150 to 170 Ma), split of major placental groups in the Early Cretaceous (100 to 130 Ma), and split of modern placental orders in the mid- to Late Cretaceous (80 to 100 Ma). The oldest fossil representatives of modern mammalian orders dated from the Paleocene and Eocene (50 to 65 Ma).

A survey of recent literature suggests that such examples are not typical and that most paleontological and molecular dates agree. This is true for intraphylum splits in many animal groups,[19, 33] the origin and divergences of major insect clades,[34] early (Paleozoic) splits among basal vertebrates[35] and tetrapods,[12, 32] and most intraordinal splits among birds[28, 29] and mammals.[32, 36, 37] Furthermore, in a comparison of 206 trees of mammals founded on molecular and morphological data,[38] congruence was commoner than noncongruence. Morphological trees were nearly twice as good as molecular trees in terms of matching between the rank orders of branching points (nodes) and oldest fossils, whereas morphological trees were 10% better than molecular trees in terms of stratigraphic consistency of the nodes. Among the molecular trees, those developed on the basis of

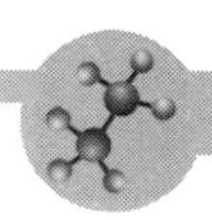

DNA or RNA data were better than those developed on the basis of protein sequences, at least in rank order of nodes and stratigraphic consistency of nodes. Protein trees, however, were best in terms of minimizing the proportion of ghost range (the postulated minimum missing fossil record implied by a tree). Fossil and molecular data are not always at odds, but both approaches have drawbacks.

Under- and Overestimating Dates

Fossils can only underestimate actual dates. Paleontologists will never find the first member of a clade, so by definition the oldest fossil must be younger than the origin of its group. Diagenesis, metamorphism, and erosion remove rocks (and included fossils) from the record and paleontologists cannot sample the earth's surface exhaustively, so much is missed.[39] Fossil occurrence may be closely correlated with the vicissitudes of rock preservation.[40, 41]

The importance of these factors has long been debated.[1] According to a pessimistic view, the fossil record is so tied to the rock record that posited mass extinctions, even the Cretaceous-Tertiary boundary (K-T) event, could be artifacts of the rock record.[41] A more optimistic view is that the K-T event and other mass extinctions are real and that the statistical manipulations used to throw doubt on them must be so crude as to be themselves doubtful.[4] Indeed, the order of fossils in the rocks is more often in agreement with the implied order of branching events in cladograms than

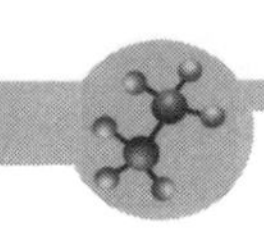

not.[42-44] These assessments have been made with the use of new age and clade metrics[42-44] that allow assessment of the reliability of fossil records and trees. The time difference between lineage divergence and the acquisition of a recognizable synapomorphy may be important biologically but unimportant geologically; disputes are measured in millions and tens of millions of years, not thousands.

It is often proposed that molecular dates are correct (with error bars) and that methods exist to correct for error.[8-10, 12] However, critics have pointed out several pervasive biases that make molecular dates too old. First, if calibration dates are too old, then all other dates estimated from them will also be too old.[22] The commonly used date for the initial divergence of the bird and mammal lines based on fossils (310 Ma) may be accurate[31] or marginally too old,[22] but other divergence dates (such as the primate-rodent at 110 Ma, arthropod-chordate at 993 Ma, fungal-metazoan at 1100 Ma, nematode-chordate at 1177 Ma, and plant-fungal-metazoan at 1576 Ma) that are commonly used[15, 18, 25, 32] are all on the basis of previous molecular studies. Some of these dates are incompatible: The nematode-chordate date (1177 Ma) cannot be older than the fungal-metazoan date (1100 Ma), because the first branching point is higher in the tree than the second. The choice of maximal dates such as these merely promulgates maximal estimates, all of which are probably too old. To use any of these dates injects circularity into the procedure, and to use

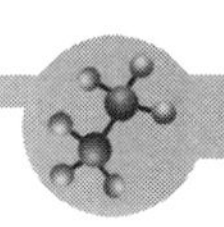

several does not help because they are not independent of each other.[14, 20]

A second biasing factor is that undetected fast-evolving genes could bias estimates of timing. Empirical and statistical studies of vertebrate sequences suggest that such non-clock-like genes may be detected and that they do not affect estimates of dating.[32] Others, however, have found that the statistical tests commonly used to exclude such sequences have unacceptably low power and could produce consistent overestimations of dates of divergence.[14, 16, 20, 24] This is because they cannot reliably reject short molecular sequences that show higher-than-normal rates of evolution, and hence the calculated time since divergence is higher than it should be. This problem may be avoided by using longer concatenated sequences and appropriate correction factors.[45]

A third source of bias relates to polymorphism. Two species often become fixed for alternative alleles that existed as a polymorphism in their ancestral species. If so, the divergence time estimated from the DNA sequences corresponds to the origin of the polymorphism, which predates the divergence of the species.[46] It is hard to judge the impact of this, but in cases of balanced polymorphisms estimated dates could be millions of years too old. Extreme cases of this are the human lymphocyte antigen and major histocompatability complex genes.[47]

A fourth biasing factor is that molecular time estimates show asymmetric distributions, with a

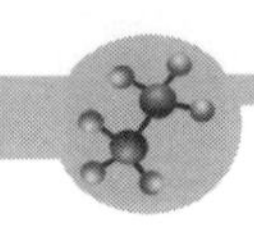

constrained younger end but an unconstrained older end. A typical plot of age estimates from different genes is right-skewed, with a large number of values at the left-hand (younger) end and a long tail of ever-older values to the right. This is because rates of evolution are constrained to be nonnegative (so the lower boundary is nonelastic), but the rates are unbounded above zero (so the upper boundary is elastic).[48] Simply taking an arithmetic mean of the estimated divergence times on the basis of all possible rates of evolution consistently overestimates the true date. This overestimation becomes more marked as the rate of molecular evolution decreases and/or the sequences become shorter. The overestimates also grow as target times become increasingly remote, so this could be a particular problem for estimates of dates in the Precambrian, for example, for the diversification of life, the plant-fungi-animals splits, and the radiation of animal phyla.[45, 48]

The common assumption that molecular dates will improve as molecular data sets become larger[8, 13, 45] may not be born out.[49, 50] Estimated dates may indeed converge, but they may converge on consistent overestimates.[48] Careful choice of genes may be a more appropriate strategy, with a focus on long and fast-evolving (yet alignable) sequences. The discrepancy between fossil and molecular dates for ancient parts of the tree of life may, however, always remain because of a combination of nonpreservation of critical early fossils and overestimation biases that cannot readily be corrected in the molecular dates.

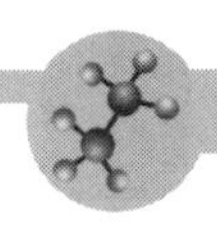

Rapprochement and Prospects

In attempting to reconstruct the single tree of life, systematists have access to three essentially independent data sets:[42-44] fossils, morphological cladograms, and molecular trees. Some parts of the tree of life are beginning to show a rapprochement as older fossils and younger molecular dates converge on a single conclusion.

A good case is the timing of the basal splits in the tree of modern mammals. The debate was polarized by rather loose statements that contrast the fossil record, where modern orders of mammals appear in the fossil record only after 65 Ma, in the Tertiary, with molecular dates that posit entirely Cretaceous (before 65 Ma) origins.[8, 10, 12] However, further analysis of the nodes in the tree has revealed that fossil and molecular evidence are in accord for 14 of the 18 mammalian orders differentiated after the end of the Cretaceous [Supporting Online Material (SOM) Text, table S1, fig. S1].

Rapprochement is to be expected; only one tree and one set of dates can be correct. But how does it happen? In the case of the ape tree, some early molecular dates were too young, and the fossil dates were too old. The paleontological error was partly a result of misclassified and missing fossils. New finds have filled the gap back to 6 to 7 Ma on the human line, but there are no fossils yet on the chimp line. In the case of the splitting of modern mammal orders, some early discussions were misinformed: Taxonomic grades were confused, and certain Cretaceous fossils were ignored. New finds have filled some gaps,[51] and other gaps are highlighted for further fossil hunting, especially in the Late Cretaceous of Africa and South America.

Are the congruent results better? Paleontological tree-making has improved methodologically since the 1960s by the widespread use now of cladistic methods. Some of the earlier disagreements followed from confused claims about identifications of fossils on the basis of sloppy character definition. Among molecular practitioners, there is a debate about whether one should use the maximum number of genes[12, 32] or select only those that may retain a strong phylogenetic signal.[16, 28] In the case of metazoan origins (Table 1), molecular dates that approach the fossil dates have been achieved more by adjusted calibration dates and different statistical filtering procedures [compare with[13, 14]] than by the use of different kinds of protein or DNA-RNA data. In the case of mammals (table S1), analyses published after 2000 seem to give more dates in agreement with fossil dates than earlier analyses, but there is no clear trend. Earlier analyses with discrepant human-rodent dates were mainly on the basis of mitochondrial DNA (MtDNA) sequencing,[52-54] but recent analyses including MtDNA genes[37, 55, 56] offer dates more in line with paleontological estimates. The changes could have as much to do with filtering and statistical processing of the data as with the choice of genes. There is no regular matching of age estimates and numbers or types of genes, but this will be a fruitful area for further consideration.

In the quest for the tree of life, it is arid to claim that either fossils or molecules are the sole arbiter of dating or of tree shape. It is more reasonable to accept that both data sets have their strengths and weaknesses and that each can then be used to assess the other.

1. M. Pagel, *Nature* 401, 877 (1999).
2. C. Darwin, *On the Origin of Species by Means of Natural Selection* (John Hurray, London, 1859).
3. A. Purvis, A. Hector, *Nature* 405, 212 (2000).
4. M. J. Benton, in *Telling the Evolutionary Time: Molecular Clocks and the Fossil Record*, P. J. Donoghue, M. P. Smith, Eds. (Taylor and Francis, London, in press).
5. C. Patterson, Ed., *Molecules and Morphology in Evolution: Conflict or Compromise?* (Cambridge Univ. Press, Cambridge, 1987).
6. A. Adoutte, S. Tillier, R. DeSalle, *Mol. Phylogenet*. Evol. 9, 331 (1998).
7. P. J. Donoghue, M. P. Smith, Eds., *Telling the Evolutionary Time: Molecular Clocks and the Fossil Record* (Taylor and Francis, London, in press).
8. S. Easteal, *Bioessays* 21, 1052 (1999).
9. G. A. Wray, *Genome Biol.* 3, 1 (2001).
10. A. B. Smith, K. J. Peterson, Ann. Rev. Ecol. Syst. 30, 65 (2002).
11. D. Graur, *Trends Ecol. Evol.* 8, 141 (1993).
12. S. B. Hedges, P. H. Parker, C. G. Sibley, S. Kumar, *Nature* 381, 226 (1996).
13. G. A. Wray, J. S. Levinton, L. H. Shapiro, *Science* 274, 568 (1996).
14. F. J. Ayala, A. Rzhetsky, F. J. Ayala, Proc. Natl. Acad. Sci. U.S.A. 95, 606 (1998).
15. N. Nikoh et al., J. Mol. Evol. 45, 97 (1997).
16. L. D. Bromham, A. Rambaut, R. Fortey, A. Cooper, D. Penny, Proc. Natl. Acad. Sci. U.S.A. 95, 12386 (1998).
17. D.-F. Feng, G. Cho, R. F. Doolittle, Proc. Natl. Acad. Sci. U.S.A. 94, 13028 (1997).
18. X. Gu, J. Mol. Evol. 47, 369 (1998).
19. M. Lynch, *Evolution* 53, 319 (1999).
20. L. D. Bromham, M. D. Hendy, Proc. R. Soc. Lond. Ser. B 267, 1041 (2000).
21. A. Cooper, R. Fortey, Trends Ecol. Evol. 13, 151 (1998).
22. M. S. Y. Lee, J. Mol. Evol. 49, 385 (1999).
23. F. Rodriguez-Trelles, R. Tarrio, F. J. Ayala, in (7).
24. P. S. Soltis, D. E. Soitis, V. Savolainen, P. R. Crane, T. G. Barraclough, Proc. Natl. Acad. Sci. U.S.A. 99, 4430 (2002).
25. D. S. Heckman et al., *Science* 293, 1129 (2001).
26. M. J. Sanderson, J. A. Doyle, Am. J. Bot. 88, 1499 (2001).
27. P. Kenrick, *Nature* 402, 358 (1999).

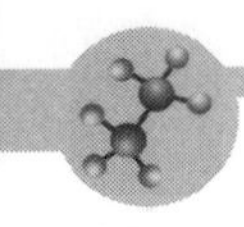

28. A. Cooper, D. Penny, *Science* 275, 1109 (1997).
29. M. Van Tuinen, S. B. Hedges, Mol. Biol. Evol. 18, 206 (2001).
30. G. J. Dyke, Geol. J. 36, 305 (2001).
31. M. J. Benton J. Mol. Evol. 30, 409 (1990).
32. S. Kumar, S. B. Hedges, *Nature* 392, 917 (1998).
33. J. Schutze et al., Proc. R. Soc. Lond. B 266, 63 (1999).
34. M. W. Gaunt, M. A. Miles, Mol. Biol. Evol. 19, 748 (2002).
35. A. Graybeal, Syst. Biol. 43, 174 (1994).
36. E. Eizirik, W. J. Murphy, S. J. O'Brien, J. Hered. 92, 212 (2001).
37. U. Arnason et al., Proc. Natl. Acad. Sci. U.S.A. 99, 8151 (2002).
38. M. J. Benton, Mol. Phylog. Evol. 9, 398 (1998).
39. D. M. Raup, Science 177, 1065 (1972).
40. A. B. Smith, Philos. Trans. R. Soc. London Set. B 356, 1 (2001).
41. S. E. Peters, M. Foote, *Nature* 416, 420 (2002).
42. M. A. Norell, M. J. Novacek, *Science* 255, 1690 (1992).
43. M. J. Benton, R. Hitchin, Hist. Biol. 12, 111 (1996).
44. M. J. Benton, M. Wills, R. Hitchin, *Nature* 403, 534 (2000).
45. M. Nei, P. Xu, G. Glazko, Proc. Natl. Acad. Sci. U.S.A. 98, 2497 (2001).
46. R. R. Hudson, J. A. Coyne, Evolution 56, 1557 (2002).
47. F. J. Ayala, *Science* 270, 1930 (1995).
48. F. Rodriguez-Trelles, R. Tarrio, F. J. Ayala, Proc. Natl. Acad. Sci. U.S.A. 99, 8112 (2002).
49. G. J. P. Naylor, W. M. Brown, Syst. Biol. 47, 61 (1998).
50. M. J. Sanderson, Mol. Biol. Evol. 19, 101 (2002).
51. Q. Ji. et al. *Nature* 416, 816 (2002).
52. S. Easteal, C. Collet, D. Betty, The Mammalian Molecular Clock (Landes, Austin, TX, 1995).
53. U. Arnason, A. Gullberg, A. Janke, X. Xu, J. Mol. Evol. 43, 650 (1996).
54. A. Janke, X. Fu, U. Arnason, Proc. Natl. Acad. Sci. U.S.A. 94, 1276 (1997).
55. S. Horai, K. Hayasaka, R. Kondo, K. Tsugane, N. Takahata, Proc. Natl. Acad. Sci. U.S.A. 92, 532 (1995).
56. P. J. Waddell, Y. Cao, M. Hasegawa, D. P. Mindell, Syst. Biol. 48, 119 (1999).
57. D. Y.-C. Wang, S. Kumar, S. B. Hedges, Proc. R. Soc. Lond. Ser. B 266, 163 (1999).
58. We thank S. B. Hedges, Z.-X. Luo, K. Padian, and anonymous reviewers for comments. Supported by the Leverhulme Trust (M.J.B.).

Supporting Online Material

www.sciencemag.org/cgi/content/full/300/5626/1698/DC1

References and Notes

Michael J. Benton (1) * and Francisco J. Ayala (2)

1. Department of Earth Sciences, University of Bristol Bristol BS8 1RJ, UK. (2) Department of Ecology and Evolutionary Biology, University of California, Irvine, CA 92697-2525, USA.

The American Museum of Natural History is spearheading an effort to create a universal tree of life. As scientists have learned new techniques and technology has advanced, the shape of the tree of life has changed. Until about 1990, most biologists thought that the tree of life had five major branches—animals, plants, and fungi at the top, protozoa and bacteria at the bottom. Then, as quoted in this article, Carl Woese, a microbiologist at the University of Illinois, revealed that "a comparison of the molecules that cells use to copy DNA and make proteins pointed to a very different tree."

This new tree has three trunks, or domains—two of microorganisms and one of more complex organisms. All three branches spring from the same ancient root. This tree is full of surprises, notes Woese: "Fungi are more closely related to animals than to plants, and reptiles have more in common with mammals than with amphibians. Starfish are closer to mammals than they are to shellfish." Only a

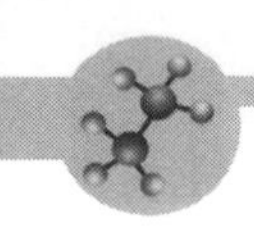

fraction of known species have been placed on this tree, but scientists believe that they now have the tools to complete it. They are confident that when they have analyzed all the data—DNA, cellular structure, anatomy, and even behavior—they will be able to tell how all known life relates. —JF

"All in the Family (American Museum of Natural History: An attempt to Trace and Record the Universal Tree of Life)"

by Thomas Hayden
***U.S. News & World Report*, June 3, 2002**

Visiting a natural history museum can be a little like attending a Broadway play; the highly polished show is the main attraction, but much of the real magic happens backstage. At the American Museum of Natural History in New York, going "backstage" means taking a trip down to the basement in a clunky freight elevator. Just beyond the carpentry workshop, in a chilled, darkened room, lurks Circe, a network of 560 desktop computers tied together with a medusa's mane of red cables. Homer's mythical enchantress was a weaver, and so is her namesake. At blinding speeds, this "cluster computer" is weaving data about living things into a fabric of relationships—a family tree of life.

When Charles Darwin published *The Origin of Species* in 1859, the tree of life was the only illustration he used. It was a rough sketch—a statement of principle,

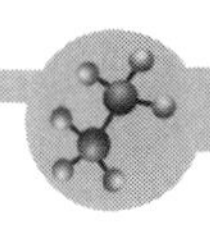

more than anything else—showing organisms branching outward as their descendants evolved into new species. Aided by a flood of genetic data and powerful computers, biologists can finally contemplate finishing Darwin's sketch. This week at the AMNH, his intellectual descendants will meet to launch a coordinated effort to assemble a universal tree of life.

It may take years or decades of collecting specimens and analyzing data. But even a partial tree will be a powerful tool. "Virtually everything we do in biology is comparative," says Joel Cracraft, AMNH's curator of birds. "You can't understand an organism without comparing it to others along the tree." It will also aid medical scientists, who will pick over the tree for clues to how diseases emerged and where cures might be found. For the rest of us, the tree already offers lessons in humility, such as humans' close kinship to the fishes and the surprising status of mushrooms, closer to us than they are to potatoes.

No sooner had Adam risen from the biblical dust than he started naming the animals; classifying nature seems to be a basic human instinct. These days the effort is called phylogeny, and it is not just an exercise in naming. Its goal is to trace evolutionary relationships, by cataloging and comparing the characters—DNA sequences, the shapes of ankle bones or seed pods—that hint at which organisms are close relatives and which have long been separated. It's like a planetwide game of "one of these things is not like the others": You put the monkeys and rabbits on one branch, the bananas and

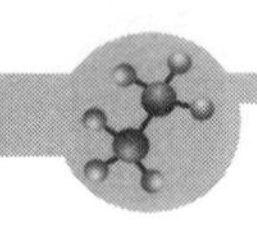

carrots on another, and pretty soon you've got the rudiments of a phylogenetic tree, or tree of life.

As scientists have learned new ways to play this game, the shape of the tree has changed, sometimes dramatically. Until about 1990, most biologists pictured a tree with five major branches—animals, plants, and fungi at the top, protozoa and bacteria at the bottom. Then University of Illinois microbiologist Carl Woese announced that a comparison of the vital molecules that cells use to copy DNA and make proteins pointed to a very different tree, with three roughly equal trunks, or domains, emerging from an ancient root.

Bacteria (think E. coli) occupy one trunk; the Eucarya—pond scum, people, and everything in between—occupy another. A third belongs to the Archaea, exotic microbes that were once lumped with bacteria but, as Woese showed, are actually a distinct domain of life, more closely related to the Eucarya. As different as people and hydrogen-eating microbes seem, "Archaea and Eucarya have the same basic tool set," says Sandra Baldauf, a University of York biologist charged with the daunting task of giving a rough outline of the universal tree at the AMNH meeting.

Climbing the branches. Molecules let Woese identify the major trunks of the tree, but specialists trying to map its luxuriant branches have found that molecules are not always enough. In the 1970s, molecular biologists were confident that they would soon work out the true tree based on differences in genetic sequences. But different genes evolve in different ways, so they often

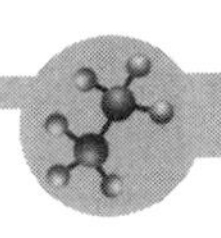

tell conflicting stories about how organisms are related. "You can't just pick a few organisms and a few genes and say, 'Here's my crazy tree,'" says Dan Janies, an AMNH researcher who along with arthropod curator Ward Wheeler developed Circe.

Physical characters, such as anatomical features and cellular structure, have their own problems. Collecting and measuring specimens is time-consuming. Appearances can deceive tree builders, as when "convergent evolution" makes lookalikes out of species that actually parted ways millions of years ago, like whales and fish or birds and bats. And in the vast domains of Bacteria and Archaea, microbes have precious little anatomy to compare.

The tree still looks scraggly, with only some 50,000 of the 1.5 million or so known species in place, but researchers think they now have the tools and know-how to fill it out. Different lines of evidence often result in contradictory trees. But when scientists analyze all the evidence—DNA, cellular structure, anatomy, and even behavior, such as nest building or seed dispersal—and include both living and fossil relatives, the data tell a more consistent tale. "Every character is part of the same story, even if they tell different versions," says Janies.

Using this "total evidence" approach, biologists are now building a convincing account of who begot whom, and they are turning up some unlikely relatives. Arrange the birds, crocodiles, and mammals on a tree, for example, and birds and mammals look like sister groups. "But if you put in the dinosaurs," says Wheeler,

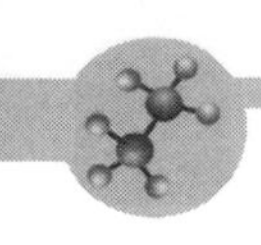

"you see the birds, crocs, and dinos on one side with the mammals far away on the other." Similar analyses, based on DNA and newly unearthed fossils, suggest that whales evolved from land animals that looked an awful lot like mangy dogs and are close cousins to the hippopotamus. "Whales are so different from other mammals that it wasn't clear where they fit," says Maureen O'Leary, an evolutionary biologist at the State University of New York-Stony Brook.

Just a few minutes tracing the existing tree is enough to shatter preconceptions. Fungi are more closely related to animals than to plants, and reptiles have more in common with mammals than with amphibians. Starfish are closer to mammals than they are to shellfish. And speaking of fish—not that oysters and starfish really are fish—Cornell University herpetologist Harry Greene has news that may unsettle those already disturbed at the thought that our ancestors were apes. In fact, says Greene, all animals with four limbs, including you, your mother, and your ape ancestors, belong to a branch of the fish lineage. "We are all fish."

Building a universal tree has practical uses, too. Scientists often try to puzzle out what a newly discovered gene does by comparing it with similar genes in other organisms—and they can sharpen those guesses if they know how closely related those organisms are. Fine branches of the tree describing how pathogens are related can help epidemiologists track the emergence and spread of diseases. And ecologists trying to predict the impact of an invasive species can often glean clues from its closest relatives in similar ecosystems.

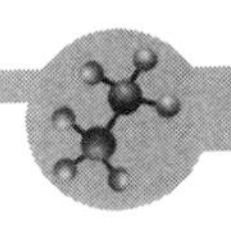

Nurturing the tree. At the meeting in New York, tree builders will lay plans to coordinate scattered efforts and graft disparate branches onto a unified tree. "We emphatically cannot go on in an idiosyncratic way," says AMNH Provost Michael Novacek. The scientists also need a truly representative sample of life to build a complete tree.

To that end, the AMNH is collecting hundreds of thousands of animal tissue samples, frozen in 10 cryogenic vats. DNA from the samples, representing every animal lineage, will be sequenced and the information fed into Circe's data-hungry maw. At the meeting, Harvard University biologist Edward Wilson also plans to push for a serious effort to catalog Earth's full diversity of life. The tree builders are likely to endorse his call. Novacek notes that as much as 90 percent of living species remain to be discovered; filling in those gaps, he says, "would provide more clues to the tree of life."

The benefits could flow back to the natural world, by helping researchers identify unique groups, with few neighboring branches, that should become the focus of special conservation efforts. Building a universal tree is the best way to understand the diversity of life. If scientists work fast enough, says Wilson, it might also be the best way to preserve it.

Searching for roots

Founders

Scientists can trace the tree back to the origin of life's three great domains, two of microbes and one of more complex creatures. But where did those lineages come from, billions of years ago? One idea is that there is no

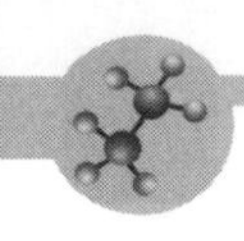

single root. Instead of one lucky organism giving rise to all others, imagine a welter of simple "cellular entities" freely swapping genes. If so, says microbiologist Norman Pace of the University of Colorado, the first modern cells probably "annealed" out of a kind of universal genetic pool.

Up the Tree

Based on comparisons of molecules and physical features, scientists are building a family tree of living things. The illustration shows key branching points along one of the three major limbs.

Archaea

These microbes, once considered bacteria, live in exotic environments like deep-sea hot springs, as well as more prosaic locales.

Bacteria

Bacteria are at least as varied as more familiar forms of life, but only a tiny fraction of them have been studied.

Eucarya

This limb encompasses organisms with cell nuclei, such as plants, animals, and fungi, which are actually closer kin to animals than to plants.

Fish and Their Relatives

One of the many branches of animals includes sharks, fish, and their relatives the tetrapods, or four-legged animals.

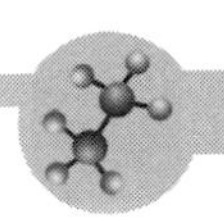

Whale Tale

Among the tetrapods are mammals, and scientists are finding surprises even here, such as the close kinship between whales and hippos.

Assembling the Tree of Life (ATOL) is a project sponsored by the National Science Foundation (NSF). NSF provides large grants to groups studying different facets of organismal relationships. Its hope is that those who earn grants will make use of modern biological analysis techniques and advanced computer information systems technology to attempt to compile a complete tree of life. This work will be performed by scientists and technical experts from fields such as DNA analysis, software design, systematics, information systems, and others.

These multidisciplinary teams will study specific problems such as the nature of particular taxa. They will also create research methods and generate databases. The NSF hopes to create a tree of life that illustrates the relationships among all known species.

Such knowledge will be applicable in a variety of areas; along with other applications,

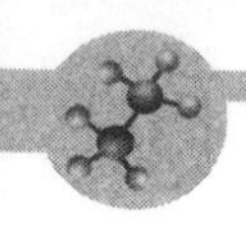

by looking at what happened in the past with closely related species in similar ecosystems, it can give ecologists a head start on figuring out what a species might do to an ecosystem to which it migrates. It can also help scientists understand related groups of bacteria or viruses, which in turn will help combat human diseases. Also, it can assist scientists who are trying to figure out what a newly discovered gene does by allowing them to compare it to similar genes in related organisms. —JF

"A New Tree of Life"
by Shanti Menon
***Discover*, June 1996**

All life, says Russell Doolittle, had a common ancestor 2 billion years ago. That's a billion years too late for most biologists.

By a billion years or so after Earth formed, life had already taken hold. Microfossils found in 3.5-billion-year-old rocks in Australia show that the first living things were prokaryotes, like today's bacteria, with DNA floating freely in their cells. For the next 3 billion years—until larger life-forms evolved—the fossil record is sparse.

Many biologists, however, believe that life split into two branches more than 3 billion years ago. That split was between bacteria and archaea—bacteria-like

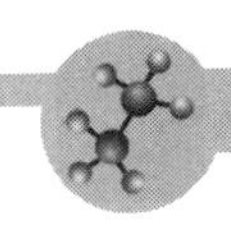

organisms that still exist today. Eukaryotes, with DNA packed in a nucleus, branched off from the archaea later and gave rise to all other life, from amoebas to people. Or so the conventional view holds. Russell Doolittle, a molecular evolutionist at the University of California at San Diego, believes that view is flawed. He has found evidence that the split between bacteria and all other life occurred much later, probably as recently as 2 billion years ago.

The standard evolutionary family tree is based on studies of mutation rates in RNA. The more similar the RNA sequences between any two creatures, the reasoning goes, the more closely related are the creatures. The problem with using RNA as an evolutionary clock, though, is that RNA sequences work within only relatively small groups of closely related organisms. Doolittle wanted a clock that he could apply on a broader scale.

He and his colleagues decided to use 57 proteins—enzymes, to be precise—found in almost every living thing. By comparing how much the amino acid sequences of these enzymes differed among assorted forms of life, the researchers could tell how closely related certain groups were, using a similar logic to the one applied in RNA studies. The researchers sampled 15 groups of organisms.

To calibrate the enzyme clock—to convert the amino acid changes, or evolutionary distance, into units of time—Doolittle and his colleagues relied on well-known species-divergence dates from the fossil record.

This enabled them to plot a straight line of the evolutionary distances they detected in enzymes versus time that extended back some 550 million years. Finally they extrapolated that line backward through time to find out when earlier lineages diverged.

All plants, animals, and fungi, according to this enzyme clock, had a common ancestor around a billion years ago. Fungi and animals shared an ancestor a little more recently than that. None of this was terribly surprising. But toward the base of the tree, things got interesting. Even after Doolittle tried to account for the slower or faster evolution characteristic of particular groups, his results still said the same thing: all living things last shared an ancestor only 2 billion years ago, instead of more than 3 billion, and the first split was between bacteria and the ancestor of eukaryotes and archaea. The archaea did not appear until 1.8 billion years ago.

Doolittle's result has attracted criticism. It's ridiculous, says Norman Pace, a microbiologist at Indiana University. The 3.5-billion-year-old Australian fossils, he says, look like cyanobacteria, a type of photosynthesizing bacteria, and the RNA family tree indicates that cyanobacteria were not the first organisms. The archaea, the eukaryotes, and most of the bacterial lineages had already been generated by the time cyanobacteria were invented, says Pace. Moreover, Doolittle's extrapolation of a mutation rate back 2 billion years, about three times the original length supported by the fossil record, strikes Pace as untenable. The evolutionary clock is not linear, he says. Evolution rates differ among different organisms, and at different times during the course of their evolution. There is simply no way you can extrapolate it. I don't know why Russ did that.

Doolittle claims that with a large data set covering a long time, all the fits and starts of evolution even out. Molecular clocks do tend to be erratic, he admits. They slow down and speed up, no doubt about it. But they never run backward. This is a challenge to people to figure out what was going on 2 billion years ago.

In the summer of 2002, the American Museum of Natural History (AMNH) in New York City hosted a scientific meeting called "Assembling the Tree of Life: Science, Relevance, and Challenges." Assembling the Tree of Life (ATOL) is a National Science Foundation–sponsored initiative that provides large grants to multidisciplinary groups to study the relationships among different organisms. The aim is to categorize all the species on Earth. This project is expected to take at least twenty years. It will require the collaboration of hundreds of scientists working in the following fields: systematics and genetics, molecular biology, information technology, and computer and software engineering.

Crucial to this project is work being carried out in all the areas that were discussed earlier in

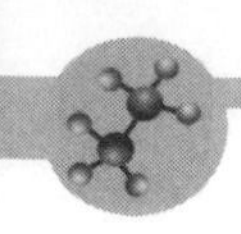

this book: constructing phylogenic trees, using DNA sequencing technology, constructing cladistic analyses, and using computer imaging and modeling techniques. The following article discusses the sweeping nature of the ATOL project and provides an overview of the specific cutting-edge research in classifying plants and animals that is currently being carried out by scientists across the continent. In addition, the article looks at some of the problems that must be addressed in order to achieve such an ambitious classification project. While some of these problems can be as basic as getting all the participants to agree on the terminology that will be used to classify the organisms, others can be more complex. For example, will the tree be designed along the lines of the traditional Linnaean "species" or use some new hierarchical design? One thing is for certain should this project achieve its goal, it will have far-reaching applications in fields as diverse as medicine, ecology, and economics, among others. —JF

From "Describing the 'Tree of Life': Attainable Goal or Stuff of Dreams?"

by Myrna Watanabe
***BioScience*, October 2002**

Will a day come when every species on Earth is described and categorized—with illustrations of type

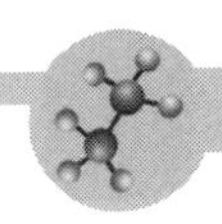

specimens and information on ecology, habitat, and other essential data, perhaps including genome sequences—all maintained in computerized files available worldwide for scientists to use and study? Famed myrmecologist and Harvard University professor E. O. Wilson described this post-space-age vision at a meeting hosted by the American Museum of Natural History (AMNH) in New York, entitled "Assembling the Tree of Life: Science, Relevance, and Challenges," held 30 May-1 June 2002. Assembling the Tree of Life (ATOL) is a National Science Foundation-sponsored initiative that will provide large grants to multidisciplinary groups studying different facets of organismal relationships . . .

Defining the ATOL program

The ATOL initiative is possibly one of the grandest projects of modern biology. It is projected to take at least two decades and will require the expertise of hundreds of scientists working in fields as diverse as systematics and phylogenetics, molecular biology, information technology, and computer and software engineering. It will require massive collaborations, which will be brought about through Tree of Life Networks (TOLNets), which concentrate on a specific empirical or theoretical problem. The networks, which may be put together to study specific taxa, create methods for research, or assemble databases, will be the basic unit for interaction among the scientists, who come from various organizations and possess diverse skills.

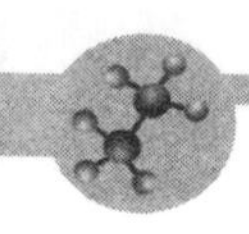

"The whole purpose of the Tree of Life program at the National Science Foundation is to do some significant clade[s] on the Tree of Life," says Diana Lipscomb, NSF's program director of systematic biology and the ATOL. The stated goal of the project is to describe the relationships among all known species, according to the report from a TOL workshop sponsored by NSF . . .

NSF's approach for ATOL is to fund large, multi-investigator, multi-institutional projects. Lipscomb gave as a hypothetical ATOL project one that considers broad questions that require a community of scientists to carry out: How are the major groups of mammals related to each other? Are carnivores related to rodents or are they related to hoofed animals? Individual species-level research, she notes, may not fit into the NSF's ATOL proposals, but such work may be funded through the agency's Systematic Biology Division, although research dollars for taxonomy and related fields are still relatively scarce. ATOL grants will be in the $2.5 million to $3 million range, as opposed to regular NSF awards, which usually are between $250,000 and $300,000, Lipscomb says.

Studying organisms under the ATOL initiative requires a background in systematics, or at least some-one on the team who is a systematist. In his remarks at the AMNH meeting, Wilson pointed out that although there are about as many taxonomists today as there were several years ago, the total percentage of scientists who are taxonomists decreased as a fraction

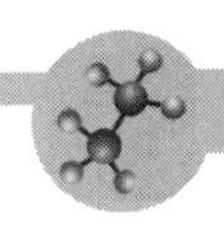

of total US scientists. "We hear all the time about taxonomists being endangered species," Yates says. "I think this effort may give rise to a whole new breed of taxonomists."

"One of the greatest challenges we face in assembling the tree of life is assembling the talent," Rita Colwell said in her ATOL talk. She added, "Systematists are as scarce as hen's teeth these days."

The ATOL program takes that into consideration. "We expect [ATOL grantees] to be training graduate students and post-docs, for example," Lipscomb says. "There are a lot of people . . . who had proposed that they would do exhibits at museums, or [include] K through 12 teachers." Furthermore, since 1996, NSF has been funding the training of taxonomists through its Partnerships in Enhancing Expertise in Taxonomy (PEET) program. Lipscomb notes, "People who are systematists are also doing the molecular biology itself."

But, as the meeting in New York shows, even absent specific NSF funding for ATOL, researchers have been engaged in work that encompasses ATOL goals. They have been building phylogenetic trees and asking serious questions about the organisms they study. Among them are John Taylor, of the University of California-Berkeley, who explained in his presentation that fungal tree of life projects not only produce phylogenetic trees but also serve as tools for other biologists. A number of fungal species have been chosen for genetic sequencing—as in the human genome project, determining the nucleotide sequence for the entire genome of the organism—which will allow comparison of fungal gene sequences with the sequences in other organisms. Such comparisons not only allow proper placement of the fungus within the tree of life but also help determine which genes are common to many taxa and which are unique to a specific taxon. Another presenter actively engaged in tree building is AMNH entomologist Ward Wheeler, who noted that recent studies on arthropod systematics show that previous classifications were "invariably wrong?" And in his presentation, Timothy Rowe, of the University of Texas-Austin, said that by identifying the genes that regulate the sense of smell in vertebrates and mapping of odor sensors through computerized tomography scans, it is possible to understand the morphology and genetics of smell in these organisms.

Bernard Wood, of George Washington University, in Washington, DC, explained how DNA sequencing studies show that the family Hominidae should include humans and chimpanzees and give a hint of when the two species diverged—about 5 million to 7 million years ago. He remarked that from the evidence, our nearest relative is *Pan* (the chimpanzee), followed by *Gorilla* and then orangutans.

Yates's work illustrates the immediate application of ATOL information to human health. Yates studies the relationship between viruses and their marine hosts. At the ATOL meeting, he discussed the problem

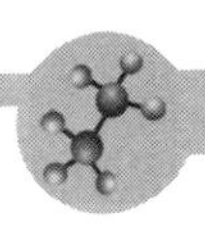

of understanding why the hantavirus carried by the deer mouse, *Peromyscus maniculatus*, was suddenly found to cause hantavirus pulmonary syndrome, which has a mortality rate greater than 50 percent. By studying preserved specimens of the mice, Yates showed that the virus had been present in the species at least since 1993. Phylogenetic analyses of both the mice and the virus showed that the hantavirus evolved with the rodent species. Yates also found that the phylogenetic tree of the virus was almost the same as that of the rodents.

More recently, he and his colleagues studied the organism that causes Bolivian hemorrhagic fever: Machupo virus, an arenavirus. Arenaviruses may cause hemorrhagic fevers in humans, and tend to be carried by rodent hosts. Although the viral reservoir is the wild mouse, *Calomys callosus*, which is found east of the Andes in Bolivia, parts of Brazil, the northern two-thirds of Paraguay, and a small corner in northern Argentina, so far only people in a small area of northern Bolivia have been infected. Yates and his colleagues found that the evolutionary lineage of the mice living in the area of human infection and carrying Machupo virus is different from that of the mice elsewhere in the range. But, Yates reported at the ATOL meeting, as logging in the Amazonian forest continues, stressing the mice and their habitat, these mice—along with the virus—are moving eastward.

Other reports at the meeting were more theoretical. Studying clades is very important, said Sean Carroll, of

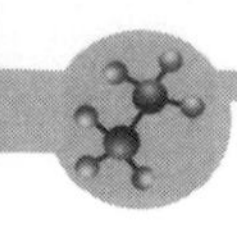

the University of Wisconsin-Madison. He explained that genes in clades are of immediate use in determining an organism's position on the tree of life. If genes are shared among members of a clade, it may be possible to infer the genomic content of the common ancestors. If genes are found in some clades and not others, it implies that, during evolution, there were episodes of expansion or loss in the species. Furthermore, Carroll said, the existence of similarly functioning (orthologous) genes may aid researchers in learning more about the phylogenetic development and morphology of the species' ancestors.

Can You Time Evolution?

As is typical in systematics, there are both questions and disagreements on interpretation of information. Norman Pace, of the University of Colorado-Boulder, pointed out in his talk that sequences are useful for determining relationships, but not for determining when a particular form evolved. Although other presenters at the meeting disagreed, Pace warned that he would question dating of age by use of sequence data. "You cannot date age by sequence change unless you have rocks," he said. "Sequence change shows adaptability."

"Different lines of descent have evolved at different rates, which confuses tree-building algorithms," Pace wrote for the meeting proceedings (to be published soon by Oxford University Press). He further noted that the tree of life as we know it is incomplete. The most abundant organisms on Earth are microbes,

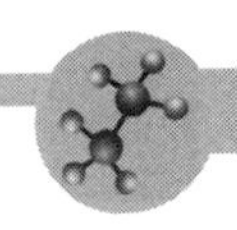

and our knowledge of microbial diversity is limited because we cannot culture about 99 percent of all existing microbes; the only way we can discover their existence is through scanning environmental samples for genes.

Pace's method for scanning microbial diversity—also used by Carl Woese, whose reorganization of the phylogenetic tree into three domains, Archaea, Bacteria, and Eucarya, still remains controversial in some quarters—is scanning environmental samples by cloning and sequencing genes found in the samples. This way, an organism that cannot be cloned can be identified by its telltale nucleic acid signs.

The Species Problem

Given the broad scope of the work already under way without support from a specific ATOL program, there are bound to be areas of question and dispute. For example, in an interview, Pace questions the use of the term *species*. He notes that the classic, Linnaean species is "a structured form. The reality of biology is messier."

NSF's Lipscomb sees the species problem as one that affects those who, like Pace, study prokaryotes. "I think that with the systematics of prokaryotes, this is probably more a real question they're wrestling with, because they don't have a view of species," she says. But for studying ATOL, "there are lots of problems that we have that you are going to need a species view of." [See original article for photo.]

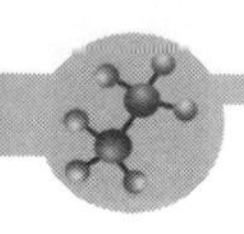

Whether all living organisms will be described or whether diversity will be quickly assessed through creation of ribosomal RNA gene libraries—as Pace does, in screening environmental microbes—may be an academic quibble among experts; NSF is the ultimate arbiter of the funds for ATOL projects. NSF is expected to announce its funding decisions on its first ATOL projects soon.

"I think that most biologists that are interested in the tree of life recognize that the main reason . . . we want the tree of life [is that] it explains why organisms are similar and why organisms are related," Lipscomb explains. "The tree of life is an information framework for biological diversity." [See original article for photo.]

The question, however, is how deeply should researchers probe relationships between known organisms. Wilson foresees an encyclopedic approach encompassing all known organisms. In his speech at the New York meeting, he promoted "a complete account of Earth's biodiversity, pole to pole, bacteria to whales, at every level of organization from genome to ecosystem, yielding as complete as possible a cause-and-effect explanation of the biosphere, and a correct and verifiable family tree for all the millions of species—in short, a unified biology." Wilson gave an example of the use of these data in a universally accessible database: He described an arachnologist pulling up online taxonomic keys; scouring online monographs, illustrations, distribution maps, and natural history records; and comparing these data

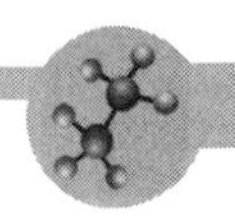

with the specimen he just collected in the field. The researcher could then enter the data from his specimen into the online database.

Wilson has support for a comprehensive database. Two members of the Social Insects Specialist Group, an advisory body of the World Conservation Union, recently complained, "Much of this information is not generally available, either because of proprietary restrictions or because of lack of coherent organization or management." They advocate creation of a free, Web-based "Biodiversity Commons" database (see Donat Agosti and Tom Moritz, "Toward the Biodiversity Commons," Species 37:9).

Pace, however, sees a difference between ATOL and Wilson's vision. "Generating a map of the tree of life is not the same as trying to make a catalog of life," he says. "I think the concept of trying to identify all the species is not a well-defined project."

ATOL-like projects

Yet whether or not researchers agree on what should or should not be studied, ATOL-like projects have taken on lives of their own. For example, in September 2000, NSF organized a workshop called "Evolution of Development Meets Tree of Life." Researchers who study evolution and development—evo-devo, for short—are interested in genes that direct development and how these are related to the tree of life. Workshop participants suggested that having information on 100 key taxa representing the major "trunks" of the tree of

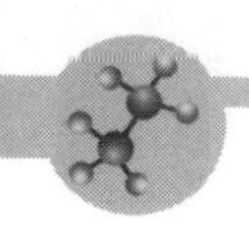

life would be essential to answer important questions about the evolutionary roles of the limited set of genes known to influence development. Both NSF and NIH's National Human Genome Research Institute support competitions specifically for organisms to be included in bacterial artificial chromosome (BAC) libraries and complementary DNA (cDNA) libraries. Organisms are recommended through submission of white papers supporting the request. A committee decides whether the organism should be given priority for library inclusion. Both BAC and cDNA libraries contain pieces of DNA from an organism that can serve as an essential tool in genomic analysis and functional analysis of genes.

Even the US Department of Energy has jumped on the tree of life bandwagon with its "Genomes to Life" (GTL) program, the stated goal of which is "a fundamental, comprehensive, and systematic understanding of life." The department, which announced grants in support of GTL in July 2002, views the program as an outgrowth of the Human Genome Project, with the goal of using sequences from microbes and higher organisms to study "the essential processes of living systems."

These programs are not aimed at solving tree of life problems solely because the technology is available and the research can be done; there are practical applications, as Yates's work on hantavirus and Machupo virus illustrates. Yates says, "I think the users [of the tree of life] will be everything from medicine, to agriculture, to economics." He noted in his ATOL talk that

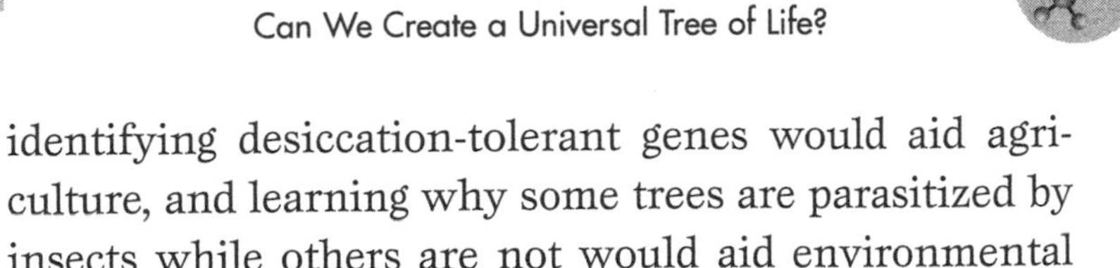

identifying desiccation-tolerant genes would aid agriculture, and learning why some trees are parasitized by insects while others are not would aid environmental studies. And ATOL information about the mosquito or avian species that harbor the West Nile virus can be put to immediate use in protecting public health.

"Nature has basically provided us with billions of years of free R&D," Yates said at the meeting. "Every one of these organisms is carrying a software package that has been tested and is functional." The goal now is to decode the software.

Web Sites

Due to the changing nature of Internet links, The Rosen Publishing Group, Inc., has developed an online list of Web sites related to the subject of this book. This site is updated regularly. Please use this link to access the list:

http://www.rosenlinks.com/cdfb/alta

For Further Reading

Dupre, John. *Humans and Other Animals.* New York, NY: Oxford University Press, 2002.

Lincoln, R. J., G. A. Boxshall, and P. F. Clark. *A Dictionary of Ecology, Evolution and Systematics.* New York, NY: Cambridge University Press, 1998.

Panchen, Alec L. *Classification, Evolution, and the Nature of Biology.* New York, NY: Cambridge University Press, 1992.

Schilthuizen, Menno. *Frogs, Flies, and Dandelions: Speciation—the Evolution of New Species.* New York, NY: Oxford University Press, 2001.

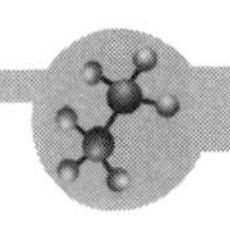

Sivarajan, V. V., and N. K. P. Robson. *Introduction to the Principles of Plant Taxonomy.* New York, NY: Cambridge University Press, 1991.

Wagner, Carl. *Character Concept in Evolutionary Biology*. Oxford, UK: Academic Press, 2002.

Bibliography

Brooks, Martin. "You Name It," *New Scientist*, Vol. 165 issue 2234, p. 8 (April. 15, 2000).

Gallagher, Richard. "The Creative Power of Naming," *The Scientist*, Vol. 1, issue 19, p. 10 (Sept. 30, 2002).

Niklas, Karl J., Michael Donoghue, and David M. Abbey. "Taxing Debate for Taxonomists," *Science*, Vol. 292, issue 5525, p. 2249 (June 22, 2001).

Pennisi, Elizabeth. "DNA and Field Data Help Plumb Evolution's Secrets," *Science*, Vol. 285, issue 5425, p. 192 (March 9, 1999).

Radford, Tim. "Metaphors and Dreams: The Paradox of the DNA Revolution Is That It Shows Us a Shining Future Without Telling Us How to Get There," *The Scientist*, Vol. 17 issue 1, pp. 24–26 (Jan. 13, 2003).

Roberts, Leslie. "Hard Choices Ahead on Biodiversity: With Many Species on the Verge of Extinction, Biologists Call for a Quick and Dirty Survey to Chart the Biodiversity of the Planet," *Science*, Vol. 241, n4874, pp. 1759–1761 (Sept 30, 1988).

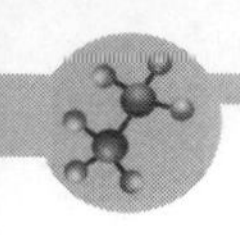

Sereno, Paul C. "The Evolution of Dinosaurs," *Science*, Vol. 284, issue 5423, p. 2137 (June 25, 1999).

Stewart, Ian. "Self-Organisation in Evolution: A Mathematical Perspective," *Philosophical Transactions of the Royal Society*, Vol. 361, p. 1101 (2003).

Stewart, Ian, Toby Elmhirst, and Jack Cohen. "Symmetry-Breaking as an Origin of Species," *Trends in Mathematics: Bifurcations, Symmetry End Patterns.*

Young, Kenneth R., Carmen Ulloa Ulloa, James L. Luteyn, and Sandra Knapp. "Plant Evolution and Endemism in Andean South America: An Introduction," *The Botanical Review*, Vol. 68, issue 1, pp. 4–21 (Jan.-Mar. 2002).

Index

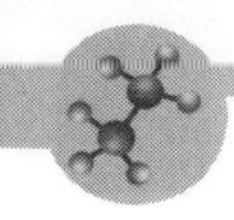

Photo Credits

Front Cover: (Top, left inset) © Inmagine.com; (bottom left) © Pixtal/Superstock; (background) © Royalty Free/Corbis; (lower right inset) © Sheila Terry/Photo Researchers, Inc.; Back Cover: (bottom inset) © Inmagine.com; (top) © Royalty Free/Corbis.

About the Editor

Jeri Freedman has a BA from Harvard University and spent fifteen years working in companies in the biomedical and high technology fields. She is the author of a number of other non-fiction books published by Rosen Publishing as well as several plays and, under the name Foxxe, is the coauthor of two science-fiction novels. She lives in Boston.

Designer: Geri Fletcher